Proceedings in Engineering Mechanics

Research, Technology and Education

This book series publishes the results of meetings dealing with material properties in engineering and science. It covers a wide range of topics, from the fundamentals of materials mechanics and applications for various industries to aspects of scientific training and career development. The volumes in the series are based typically on primary research materials presented at conferences, workshops, and similar scientific meetings, and represent comprehensive scientific and technical studies.

More information about this series at https://link.springer.com/bookseries/16733

Lucas F. M. da Silva · Paulo A. F. Martins ·
Uwe Reisgen
Editors

2nd International Conference on Advanced Joining Processes (AJP 2021)

Selected Contributions of AJP 2021

Springer

Editors
Lucas F. M. da Silva
Faculty of Engineering
University of Porto
Porto, Portugal

Paulo A. F. Martins
Instituto Superior Técnico, IDMEC
University of Lisbon
Lisbon, Portugal

Uwe Reisgen
ISF
RWTH Aachen
Aachen, Germany

ISSN 2731-0221 ISSN 2731-023X (electronic)
Proceedings in Engineering Mechanics
ISBN 978-3-030-95462-8 ISBN 978-3-030-95463-5 (eBook)
https://doi.org/10.1007/978-3-030-95463-5

This Springer imprint is published by the registered company Springer Nature Switzerland AG
The registered company address is: Gewerbestrasse 11, 6330 Cham, Switzerland

Preface

This volume of *Proceedings in Engineering Mechanics—Research, Technology and Education* contains selected papers presented at the 2nd International Conference on Advanced Joining Processes 2021 (AJP 2021), held in Sintra (Portugal) during 21–22 October 2021 (www.fe.up.pt/ajp2021).

The goal of the conference was to provide a unique opportunity to exchange information, present the latest results as well as to discuss issues relevant to advanced methods of joining such as friction stir welding, joining by plastic deformation, laser welding, advanced mechanical joining, adhesive bonding, hybrid joining, etc. The focus is on process optimization in experimental and simulation terms, metallurgical and material behavior associated with joining, engineering properties and assessment of joints, health and safety aspects of joining, durability of joints in service, industrial applications and education.

Approximately 170 papers were presented by researchers from more than 20 countries. In order to disseminate the work presented in AJP 2021, selected papers were prepared which resulted in the present volume dedicated to advanced joining processes. A wide range of topics are covered resulting in 12 papers dealing with the most recent research topics concerning mechanical joining (first section), welding (second section) and adhesive bonding (third section). The book is a state-of-the-art of advanced methods of joining and also serves as a reference volume for researchers and graduate students working with advanced joining processes.

The organizer and editor wish to thank all the authors for their participation and cooperation, which made this volume possible. Finally, I would like to thank the team of Springer-Verlag, especially Dr. Christoph Baumann and Ute Heuser, for their excellent cooperation during the preparation of this volume.

Paulo Martins would like to acknowledge the support provided by Fundação para a Ciência e a Tecnologia of Portugal and IDMEC under LAETA-UIDB/50022/2020 and PTDC/EME-EME/0949/2020.

Porto, Portugal
Lisbon, Portugal
Aachen, Germany
November 2021

Lucas F. M. da Silva
Paulo A. F. Martins
Uwe Reisgen

Acknowledgements Paulo Martins would like to acknowledge the support provided by Fundação para a Ciência e a Tecnologia of Portugal and IDMEC under LAETA-UIDB/50022/2020 and PTDC/EME-EME/0949/2020.

Contents

Mechanical Joining

Load Bearing Behaviour of Thermoplastic Composite/Metal Hollow Structures with Multiscale Form Closure

Raik Grützner, Veit Würfel, Roland Müller, and Maik Gude

Abstract Joining systems with a multi-scale form closure function enable to pass very high loads into rod- and tube-shaped fibre reinforced structures and achieve high degrees of material utilization for the composite part. The multi-scale load application is based on a combination of macro and meso form closure with adaptation to the fibre angle of the braided preform. The design of the form closure elements on different scales determines the joint failure mode, which can be cohesive or adhesive failure. This paper presents the investigation of the influence of a multi-scale form closure on the load bearing capability of hybrid joints made of carbon fibre reinforced thermoplastic and aluminium under static tension load. Using structurally relevant combinations of meso structures and macro contours, numerical sensitivity analyses are used to determine the relationships between geometric parameters of the form closure elements, the transmittable forces and the failure behaviour. Experimental investigations of the joint strength of meso, macro and combined-structured tubular specimens show the beneficial effect of multi-scale structuring to increase the joint strength and match with the numerical simulations. Concluding advises for the contour joints design are given.

Keywords Multiscale form closure · Contour joints · Joint strength · Failure behaviour · Numerical analyses · Friction-based contact · Pull-out test

---

R. Grützner (✉) · R. Müller
Fraunhofer-Institute for Machine Tools and Forming Technology IWU, Reichenhainer Straße 88, 09126 Chemnitz, Germany
e-mail: raik.gruetzner@iwu.fraunhofer.de

R. Müller
e-mail: roland.mueller@iwu.fraunhofer.de

V. Würfel · M. Gude
Technische Universität Dresden, Institute of Lightweight Engineering and Polymer Technology, Holbeinstr. 3, 01307 Dresden, Germany
e-mail: veit.wuerfel@tu-dresden.de

M. Gude
e-mail: maik.gude@tu-dresden.de

L. F. M. da Silva et al. (eds.), *2nd International Conference on Advanced Joining Processes (AJP 2021)*, Proceedings in Engineering Mechanics,
https://doi.org/10.1007/978-3-030-95463-5_1

1 Introduction

For the production of lightweight profiles like shafts, tubes or tension–compression struts, hollow structures with a closed cross-section are often used having a more favourable stiffness-to-weight ratio than corresponding solid material variants. The lightweight potential can be enhanced further by the use of anisotropy in composite materials because of the highly orientated loads within these structures. Areas with high tribological stress or high functional density are designed in the sense of multi-material design using metallic connection and fastening elements (bearing seats, retaining rings, shaft nuts, etc.) in order to be able to operate gears and bearing elements. A major challenge is the introduction of forces from the metallic component into the laminate, as the forces must be transmitted from the surface of the fibre reinforced plastic (FRP) hollow structure to the fibres via interlaminar shear within the matrix. In established methods for joining the metallic load introduction elements like bonding, press-fitting or joining using bolt and pin elements the connection is formed after the manufacture of each single component in a subsequent process step. Often times this is related to labour-intensive operations which add extra costs and may induce damage into the laminate during manufacture (Fleischer et al. 2021). A novel manufacturing approach in which hydroformed and multi-scale structured metallic load introduction (LI)-elements are intrinsically joined with thermoplastic tape-braided preforms in an integral bladder-assisted moulding process has been presented (Würfel et al. 2020). In this work, the load bearing behaviour of thermoplastic composite/metal hollow structures with a multi-scale form closure is studied. The performance these hybrid joints with metallic LI-elements, which are structured on different scales, are investigated numerically and experimentally in quasi-static tensile tests. A numerical model that predicts the failure behaviour of the hybrid structure is presented. Numerical sensitivity analyses are used to determine the relationships between geometric parameters of the form closure elements, the transmittable forces and the failure behaviour of the hybrid structure with the aim to optimize future designs.

Contour joints prove to be able to transfer high mechanical loads in combination with a high material utilization rate (Hufenbach et al. 2008). The design concept and its capabilities has been studied for FRP hydraulic cylinders (Hufenbach et al. 2006) and for composite drive shafts (Gude et al. 2014). In an integral bladder-assisted moulding process (IBM) consolidation and simultaneous joining of a braided hybrid commingled yarn preform into a metallic functional element is done in one single step. A form closure on the macro level was investigated, whereby no additional strengthening by micro or meso structuring on the surface was studied (Hufenbach et al. 2008). Structuring on the micro scale via laser subtraction processes was investigated to either increase the interface area or create interlocking elements with the result of an increase in bonding strength (Baburaj et al. 2007; Wang et al. 2016). Alternatively, various surface treatments improve the joint strength of intrinsic hybrid composite structures (Gebhardt and Fleischer 2014; Zinn et al. 2018). In other studies aluminium carbon fibre reinforced plastic (CFRP) hybrid hollow structures were intrinsically manufactured and joined by rotational moulding (Fleischer et al. 2015; Nieschlag et al. 2018), in which a braided preform is impregnated by thermoset resin. Intrinsically joined tubular structures consisting of multi-scale structured aluminium LI-elements and carbon fibre reinforced thermoplastic (CFRTP) hollow structures were presented in Würfel et al. (2020), Gude et al. (2021). The utilisation of hydroformed metallic LI-elements and braided thermoplastic tapes in combination with the bladder-assisted moulding process enables a large-scale production of hybrid composite-metal structures without the need for any further trimming or joining operations. In two separate process paths two semi-finished parts are manufactured. A multi-layer unidirectional thermoplastic tape-preform was produced by thermoplastic tape braiding (Garthaus et al. 2015) and the metallic LI-element is manufactured in a two-step hydroforming process in which the structuring on meso scale and the desired macro contour for a form closure connection is formed (Grützner et al. 2021). Both semi-finished parts are intrinsically joined in the IBM in which the tubular braided preform is formed into the structured LI-element by applying pressure to the internal bladder when the melting temperature of the matrix material is reached. Consequently, the composite is moulded along the contour of the structured LI-element creating a form closure joint (Barfuss et al. 2018). A multi-scale form closure combination of meso structuring and macro contouring improved the quasi-static pull-out strength of the joint compared to sole meso and macro contouring (Gude et al. 2021).

The prediction of internal material structure of the CFRTP and the metal joining partner in intrinsic hybrids in terms of cracks and stresses using finite-element-methods accelerates research and development. For the calculation of the homogenised mechanical properties of an intrinsic hybrid composite with a form closure, representative volume elements (RVE) with material models based on connected rheological elements and an elastic–plastic damage material model for the interface is used in Kießling et al. (2017). The sensitivity to geometrical design variables and asymmetrical stackings of fibre metal hybrid shell structures is investigated by a simplified two-dimensional finite element model and constant stress exposure method for material degradation in Herwig et al. (2021). In Nieschlag et al. (2021) a design of experiment (DoE) study was used to numerically determine the influence of different design parameters of the joint. The maximum principal stresses in the modelled adhesive layer were selected and analysed due to their main responsibility for joint failure. For tubular intrinsic hybrid composite structures a numerical model was developed, to study the influence micro structure on the overall strength and failure behaviour of the metal-composite interface (Hirsch and Kästner 2017). In Barfuss et al. (2016) a design of experiment (DoE) approach to optimise the macro contour is presented. Multiple FE simulations with linear elastic material behaviour and the CUNTZE failure mode criterion were used in the pareto optimisation. The meso structure was simplified with the use of a cohesive zone model.

Because most studies on intrinsic tubular hybrid joints focused on the experimental results or did not calculate a multi-scale form closure joint on all relevant scales a simulative approach to predict the failure of these structures is presented.

The load bearing capability of meso and macro structured hybrid tubular joints with a form closure is numerically investigated. An "elastic damage" model based on degradation theory is used to calculate the damage propagation during pull-out tests. The numerical results are compared with experimental quasi-static tensile tests with different structuring configurations.

2 Materials and Methods

2.1 Design and Geometry

In order to transfer high mechanical loads in CFRP/metal hybrid contour joints information about several mechanical and physical properties of the materials and manufacturing process are required. The load bearing capability is governed by the structure of the LI-element as well as the failure behaviour of the chosen material combination of aluminium and carbon fibre reinforced polyamide 6 (PA6). The LI area has a fibre appropriate design concerning a homogeneous load transfer from the metallic LI-element into the hollow CFRP part. Surface structures on different scale levels from meso to macro are analysed with a discretization of the structuring levels defined as smaller than 1000 μm for the meso level and higher than 1000 μm for the macro level. In this work, a rhombus design adapted to the braided fibre architecture and circumferential undercuts were realised as mesoscopic structuring with and without a macro contour (Fig. 1). The rhombus is designed to interact with the whole tape and adapted to the thermoplastic tape width of 3 mm and braiding angles of the outer plies of the CFRP component. The circumferential grooves are designed to induce undulations on the meso scale in the laminate in which the resultant form closure of the laminate is oriented in load direction. The specimens interface area between the LI-element and CFRP part has an diameter of 46 mm. The length of the meso, macro and combined form closure combination is 30 mm. The design of the tapered macro contour is investigated in Gude et al. (2021).

2.2 Thermoplastic Tape-Braided Preform

The investigated tape-braided preform is made of Celanese, CF-Polyamide 6 Celstran® CFR-TP PA6 CF60-03 slit tape material. The tape is slit to a width of 3 mm for textile manufacturability in the braiding machine. The fibre volume fraction is 48% and it contains T700S carbon fibres from Toray Inc. The BASF PA6 Ultramid B24 N03 matrix system is used, which is characterized by low viscosity and thereby favourable processibility (Barfuss et al. 2016). The material is braided using a tubular steel mandrel with an outer diameter of 38 mm and a Herzog KFh/48-100 braiding machine. For the experiments in this investigation,

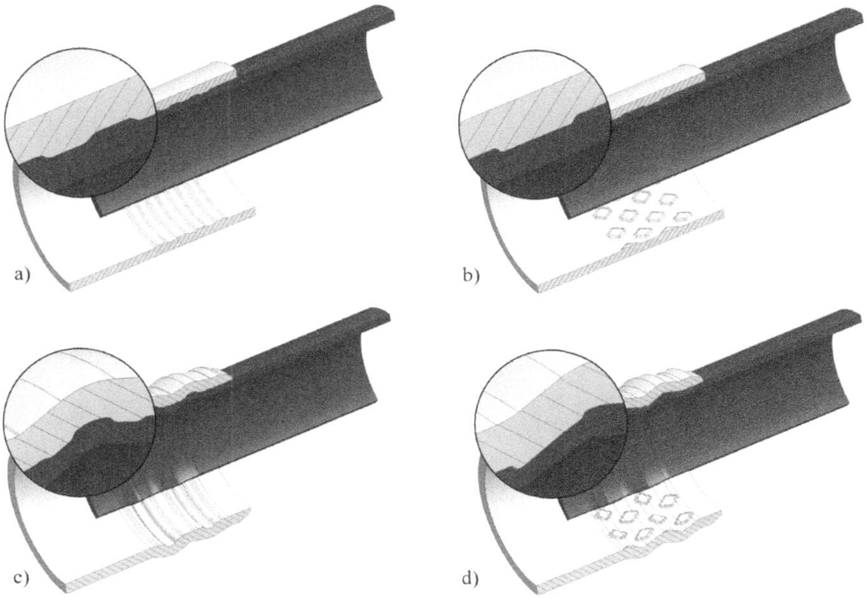

Fig. 1 Section view of the form closure combinations evaluated in this work: **a** meso groove, **b** meso rhomb, **c** meso-macro groove, **d** meso-macro rhomb

layups of [$\pm 30°_2/0°_7/\pm 30°_2$] are manufactured with a 2 × 2 biaxial twill fibre architecture in the braided plies. Unidirectional 0°-plies are added to reinforce the laminate in the main load direction. The layup is optimized to match the CTE of the aluminium load introduction elements in order to minimize residual stresses during manufacturing (Barfuss et al. 2016). An additional braided [$\pm 70°_5$]-bandage is used to reinforce the laminate in the structured area to secure the form closure against transverse contraction of the tubular structure when tensile load is applied. The intersection points of the textile improve preform stability but also restrict the forming capability of the structure. Therefore, a novel manufacturing strategy to implement the reinforcing bandage has been developed (Würfel et al. 2018).

2.3 *Two-Step Hydroformed Load Introduction Elements*

In order to manufacture metallic LI-elements with a multi-scale structure, which is scalable for industrial application, a novel two-phase hydroforming process has been developed. It consists out of a combination of external and internal hydroforming steps to shape the desired meso and macro structures on the LI element. During the process, a tubular aluminium profile is inserted into a macro contoured outer tool together with the meso structured inner tool. The process as well as its

Fig. 2 Cross sections of hydroformed load introduction elements: **a** meso groove, **b** meso rhomb, **c** meso-macro groove, **d** meso-macro rhomb

possibilities and limitations has been extensively covered in Grützner et al. (2021). Within the scope of this investigation, the aluminium alloy EN AW-6060 is examined for the metallic LI-element. The best cold formability is achieved in the stabilised delivery condition T4 which is solution annealed and naturally aged. For this reason, the forming production of LI-elements with meso and macro structuring is carried out in the T4 condition. The hydroformed specimen are shown in Fig. 2.

2.4 Integral Bladder-Assisted Moulding Process

In order to evaluate the potential of the load bearing behaviour of thermoplastic composite/metal hollow structures with multiscale form closure specimen are manufactured in a convection oven in the integral bladder-assisted moulding process. The tape-braided preforms and the hydroformed aluminium LI-elements are put into an outer moulding tool in which the LI-element serves as the tool itself. The specimens are heated up to the processing temperature of 230 °C with a heating rate of 10 K/min. After a 15 min isothermal consolidation dwell time and a pressure application of 6 bar the specimen is cooled down to room temperature with a cooling rate of 10 K/min. Both semi-finished parts are joined in the integral bladder-assisted moulding process in which the CFRP part is simultaneously consolidated and formed into the metallic LI element.

2.5 Numerical Approach

2.5.1 Structural Simulation

For the numerical evaluation of the hybrid composite joint, the solver LS-Dyna 11.0 of Livermore Software Technology Corporation (LSTC) is used, as it provides beneficial material models for the representation of FRP and has found wide use in the computational calculation of FRP structures.

2.5.2 Material Models

Braided thermoplastic composite

In LS-Dyna, specific material models are available to describe the fibre composite properties. For the simulation of braided structures, especially crash absorbers, the material model *MAT_LAMINATED_COMPOSITE_FABRIC, (*MAT_058) shows the best results (Schweizerhof et al. 1998; Xiao 2009; Xiao et al. 2009; Matheis 2013; Jackson et al. 2014). *Mat_058 has a nonlinear anisotropic material law. Depending on the type of failure surface, this material model can be used to model composites with unidirectional layers, complete laminates and fabrics.

Based on the work of *Hashin* (Hashin 1980), *Matzenmiller* (Matzenmiller et al. 1995) developed a fundamental model to describe the elastic-brittle behaviour of FRP. The so-called "Elastic Damage" model is based on the degradation theory and assumes a macroscopic approach. The assumption is made that micro-cracks propagate in the material structure during deformation, which lead to a continuous decrease in the stiffness properties. To describe this damage behaviour, LS-Dyna calculates so-called damage variables ω_i for different load conditions.

The following equation describes the dependence of the damage variables on the calculated strain ε_i and the maximum permissible elastic strain ε_f, which are used as input variables for the calculation.

$$\omega_i = 1 - e^{-\frac{1}{m_i e}\left(\frac{\varepsilon_i}{\varepsilon_f}\right)^{m_i}} \tag{1}$$

The parameter m_i of the damage exponent results from the strain ε_q reached at maximum strength:

$$m_i = \frac{1}{ln\left(\frac{\varepsilon_q}{\varepsilon_f}\right)} \tag{2}$$

The damage variable ω_i can only take values between zero and one. If there is no material strain, the exponential expression corresponds to the value one and the equation results in the value zero. As the damage to the material structure progresses, ω_i approaches the value one.

In *MAT_058 a so-called "Postdamage" parameter can be set. This value (SLIMx) specifies a residual load-bearing capability after the failure of a layer. The following relationship exists:

$$\sigma_{max} = SLIM \times strength \tag{3}$$

In this case "strength" means the respective strength limit. The residual load-bearing capabilities are defined separately for the load types tension, compression and shear for fibre and matrix, whereby the adjustable value range is between one and zero (Livermore Software Technology Corporation 2019). The utilised mechanical material properties for the carbon fibres, the polymer and the UD tape material are listed in and taken from Barfuss et al. (2016).

Aluminium material

The elasto-plastic material model *MAT PIECEWISE LINEAR PLASTICITY (*MAT_24) is used to model the aluminium LI-elements. Here, an arbitrary stress–strain curve and a strain rate dependence can be defined to describe the material behaviour. Furthermore, a failure can be specified based on a plastic strain or a minimum time step size (Livermore Software Technology Corporation 2019). With the aim of being able to introduce and transfer high loads into the aluminium LI-element, the flow curve in the state T6 is used in the structural simulations. Failure is not defined. For the numerical investigations of the joint design and the forming of the multi-scale structures, the aluminium states T4 and T6 used are characterised at room temperature in a quasi-static tensile test according to DIN EN ISO 6892–1 using five tubular samples each. Table 1 lists the characteristic values determined.

Table 1 Material properties EN AW-6060 in the states T4 and T6

Parameter	Unit	EN AW-6060-T4	EN AW-6060-T6
Density ρ	g/cm^3	2.75[b]	2.75[b]
Wall thickness t	mm	2.0	2.0
Young's modulus E	GPa	68	69
Yield strength $R_{p0,2}$	MPa	65	188
Tensile strength R_m	MPa	151	227
Breaking elongation $A_{100\ mm}$	%	13	6
CTE α[a]	10^{-6} K^{-1}	23.4[b]	23.4[b]

[a] 20–100° C

[b] Manufacturer's information

2.5.3 Model Structure and Boundary Conditions

A modelling strategy consisting of a combined stacked-shell and layered-shell method is used to model the laminate structure. The contour is mapped with constant radii for all layers. For the specific analysis of each individual layer, the use of thick shell elements is resorted to. The thick shell element is an 8-node element based on the shell theory, but with an additional strain component described by the element thickness (Hallquist 2019). By stacking thick shell elements, any laminate structure can be discredited even for small radii of a contour joint. With the focus on the investigation of influencing parameters for the design of contour joints by means of fast variant simulations, no delamination processes are considered for simplification, i.e. all layers are firmly connected to each other. In the model, this is realised by using identical node numbers of adjacent elements of the respective layers in the laminate. By dispensing with cohesive or contact elements between the layers, short calculation times can be achieved and thus the additional effort due to the use of thick shell elements can be compensated.

The basic FEM model of the pull-out test corresponds in geometric dimensions and layer structure to the macro-contoured real specimens with circumferential stiffening bandages (see Fig. 3).

To reduce the computational effort, the model is reduced to a 45° sector with cyclic boundary conditions. At the intersecting edges of the sector, the degrees of freedom are each connected in such a way that the same behaviour is repeatedly enforced cyclically. The total of 25 fibre composite layers with the fibre orientations $[\pm30°_2/0°_7/\pm30°_2/\pm70°_5]$ are combined to 12 layers of thick shell elements by mapping with two integration points over the element thickness. An exception are the middle UD layers 7–9, which are represented by three integration points in layer

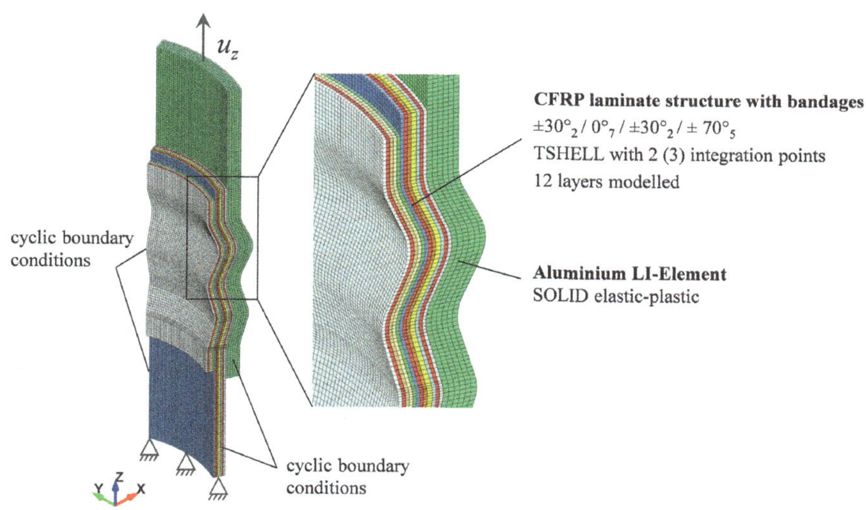

Fig. 3 Model structure of the pull-out simulation with a tapered macro contour

4. The sum of the individual layers results in a wall thickness of the fibre composite of 2.1 mm or 3.5 mm in the connection area. Undulation of the tapes in the braided layers is not taken into account.

The aluminium load introduction element is modelled with five SOLID elements over the wall thickness of 2.5 mm. The element formulation ELFORM 2 is used, as this represents a fully integrated higher order element and thus does not require an hourglass control.

The load is applied to the edge nodes of the aluminium tube via a displacement $u_z = 5.0$ mm. All degrees of freedom are bound, except for the translation in the Z-direction. In the calmping area of the composite, all node translations and rotations $u_x = u_y = u_z = rot_x = rot_y = rot_z = 0$ are set. For the analysis of the force–displacement curves, the reaction forces of the clamped nodes are recorded in total and output in the result file with the corresponding time step.

In Fig. 4 the model structure of the different meso and macro form closure configurations is shown.

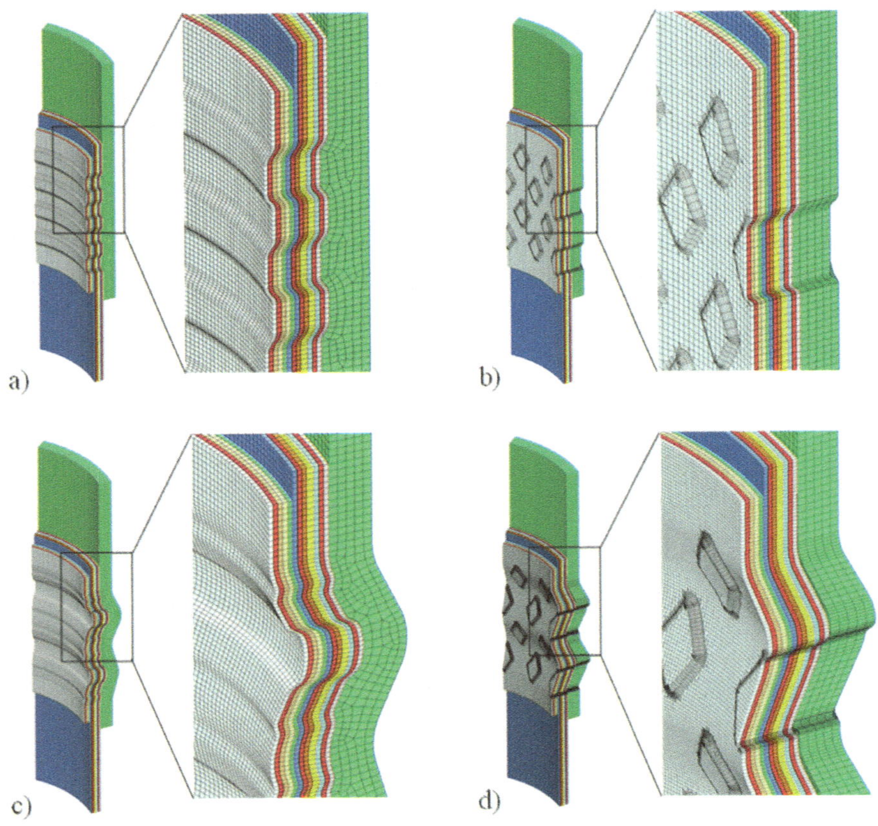

Fig. 4 Model structure of the pull-out simulation, **a** meso groove, **b** meso rhomb, **c** meso-macro groove, **d** meso-macro rhomb

2.5.4 Contact Definition

In order to describe interactions in the interface between CFRP and aluminium, a friction-based contact approach is pursued in this work. Coulomb's friction law is applied, in which the quotient of the normal force F_N and the frictional force F_R yields an average, constant friction coefficient μ. A distinction is made between static and dynamic friction. The exponential decay friction model proposed by Oden and Martins (1985) is implemented in LS-Dyna to describe the boundary surface friction property. The literature does not provide any friction coefficients for the material pairing of fibre-reinforced PA6 and aluminium. A reference point is provided in Schürmann (2007) for the friction pairing FRP/steel with a friction coefficient $\mu = 0.1$–0.2. Since the frictional force F_R is directly proportional to the normal force F_N between the friction surfaces, the mechanical connection properties and the failure behaviour are significantly influenced by the friction coefficient. In the modelling strategy based purely on form closure, no frictional components are taken into account in the connection. Thus, no normal force acts on the joining partners at the beginning of the pull-out. For this reason, only the coefficient of sliding friction is varied in the simulations and experimentally validated. Over 30 different contact formulations are provided in LS-Dyna to describe interactions between the individual elements.

The contact *CONTACT_AUTOMATIC_SURFACE_TO_SURFACE, chosen because of its robustness, is used to check all elements of the fibre composite for contact with elements of the metallic partner in order to describe their interaction and prevent penetration. LS-Dyna uses the penalty method, in which linear spring elements are inserted between the penetrating nodes and the associated contact area to compensate for overlaps. The spring stiffness matrices are then incorporated into the overall stiffness matrices. The contact forces are calculated based on the penetrations of the slave or master nodes to the associated contact segment and assigned to the nodes in contact.

2.5.5 Evaluation Methods

To characterise the stiffness and load-bearing behaviour of the hybrid joints, quasi-static pull-out tests are carried out. Conditioning of all samples according to DIN EN ISO 1110 before testing ensures the reproducibility of the polyamide properties. Specimen testing was performed under quasi-static tensile load at an INSTRON servo-hydraulic testing machine. Loading was applied by a constant speed of 0.5 mm/s. The deformation and displacement of the joining partners in relation to each other are recorded during the tensile test via a stochastically distributed black-and-white pattern using optical measurement technology (GOM ARAMIS 3D) and the displacement is calculated according to the principle of digital image correlation (DIC), see Fig. 5. The specimens are inspected visually

Fig. 5 Displacement measurement during tensile testing with GOM ARAMIS 3D resolution: 4096 × 3072 pixels, measuring volume (L × W × D): 150 × 115 × 95 mm, frame rate: 15 Hz

after testing. The length of the structuring as well as the inner diameter of the aluminium LI-element are used for the calculation of the joint strength. Therefore the length increase due to the meso or macro structuring is not taken into account.

3 Results

3.1 Quasi-static Tensile Tests

To investigate the strenghtening effect of a multi-scale structure, the meso structures and the macro contour with two undercuts as well as their combinations were examined and the load-bearing and failure behaviour compared. All samples fail uniformly due to fibre fracture of the laminate as a result of excessive tangential compressive stresses in the undercuts. In earlier investigations with the use of in-situ computed tomography analysis it was shown that as a result of the axial tensile load, a significant tapering of the FRP cross-section and an associated detachment of the positive locking elements occurs (Gude et al. 2021). The comparison of strength and pull-out force as well as the fractured specimens is shown in Fig. 6. The specimens with a sole macro contour have the lowest strength compared to the meso or multi-scale structured specimen. The tests indicate that the combination of meso and macro form closure increased the maximum force and strength of the joint compared to a sole meso or macro form closure. The determined joint strengths of the multi-scale form closure joints are 11 MPa for two undercuts and

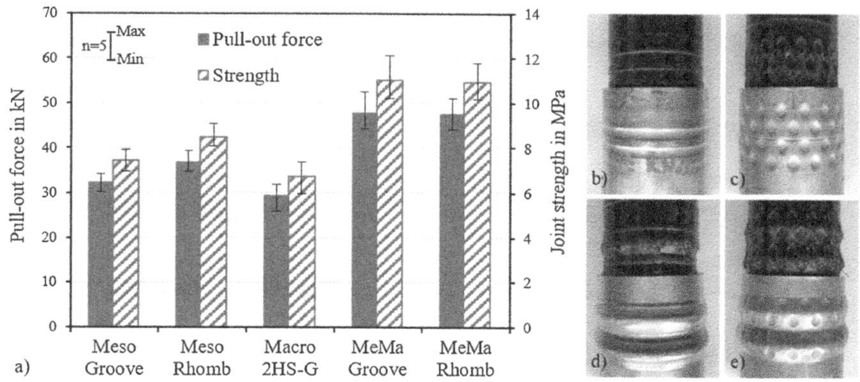

Fig. 6 Results of testing: **a** maximum force and joint strength in the pull-out test for the meso structures groove and rhombus, the macro contour and the combined structured joints as well as the specimen failure of **b** meso groove, **c** meso rhomb, **d** meso-macro groove, **e** meso-macro rhomb

30 mm structuring length and achieve equivalent shear strengths of adhesive bonds. The force–displacement behaviour is compared to the numerical results in Sect. 3.2.3.

3.2 Numerical Simulation

3.2.1 Form Closure on the Macro Scale

In the first step, the pull-out simulations are carried out with the contour geometries presented on a macro scale. Figure 7 shows an exemplary force–displacement curve and longitudinal sections of the FE model with representation of the fracture criteria fibre fracture and inter-fibre fracture in the CFRP at discrete times or pull-out distances. After 0.75 mm displacement (point 1), damage is evident in the $\pm 70°$-bandages of the laminate, especially in the force-transmitting flanks of the shaft contour. At the force maximum (point 2), the first layers of the bandages fail due to fibre and matrix breakage, while the UD layers oriented in the pull-out direction remain undamaged as expected. In the further course of the pull-out (points 3 and 4), a complete failure occurs due to fibre fracture of both the inner and outer braided layers. This reduces the resistance to the cross-sectional taper of the CFRP and leads to a total loss of the form closure. This failure behaviour coincides with the experimental results.

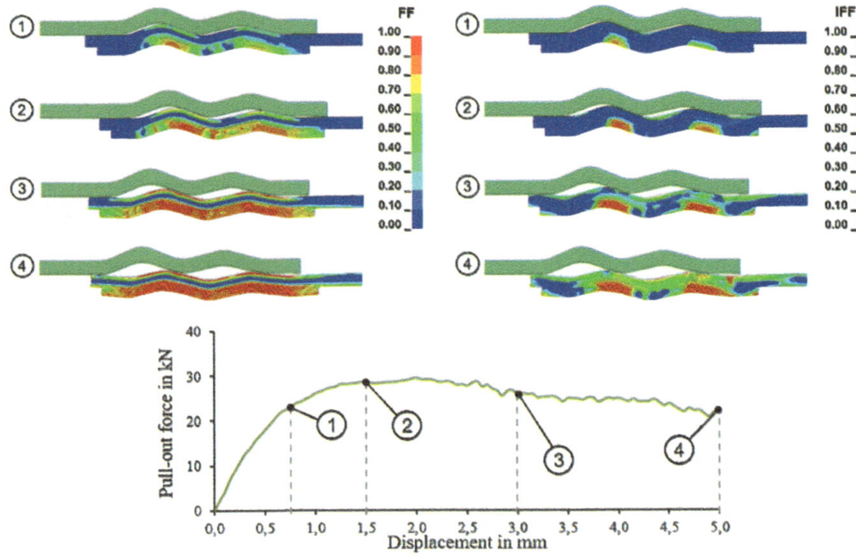

Fig. 7 FE model of the pull-out simulation with fracture criterion for fibre fracture (FF) and inter-fibre fracture (IFF) after 0.75 mm (1); 1.5 mm (2); 3.0 mm (3) and 5.0 mm (4) pull-out displacement as well as force–displacement curve of the joint with macro form closure

The analysis of the failure-relevant stresses in the individual fibre layers at the force maximum shows almost equally distributed tensile and compressive stresses in the outer braided layers and predominantly tensile stresses in the UD layers. Due to the fibre orientation, compressive stresses dominate in the bandages both in fibre direction and transverse to the fibre direction, see Fig. 8. The validation of the FE base model with a macro contour is done by comparing measured and calculated force–displacement curves as well as by characterising the damaged areas after the pull-out test.

Figure 9 shows the force–displacement curves simulated with different friction coefficients. In addition, the range of minimum and maximum force–displacement curves of the experimental test results is plotted. A very high agreement of the curves over the complete pull-out displacement as well as the sensitivity of the friction coefficient is evident. With a coefficient of friction of $\mu = 0.1$, the experimental force–displacement curve coincides with the calculations, while a lower coefficient of friction is underestimated and a higher one is overestimated. This shows the suitability in principle of the selected material models and the friction-based contact approach in the interface of the joining partners for calculating the mechanical structural properties. The determined model parameters form the basis for all further numerical investigations.

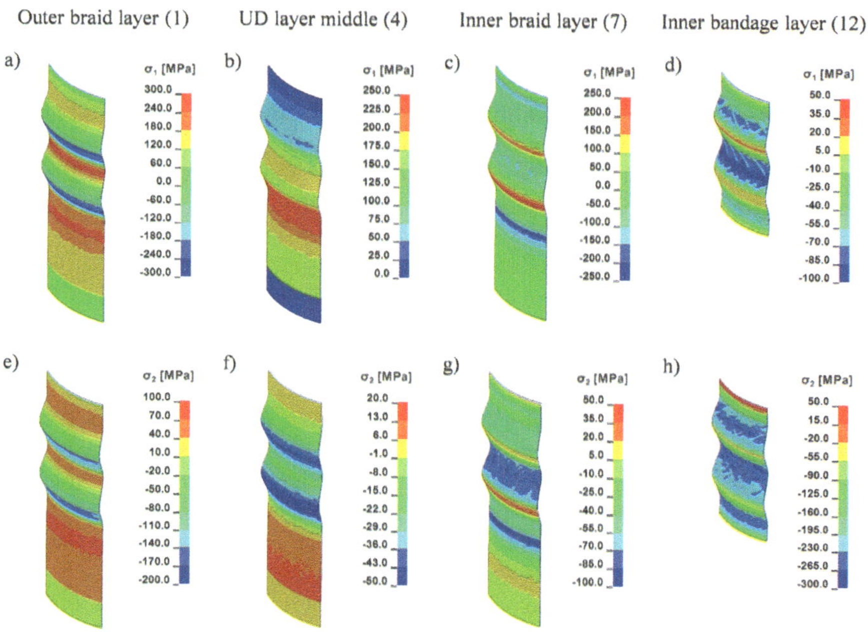

Fig. 8 Stress in fibre direction **a–d** and in transverse direction **e–h** at maximum force for different layers

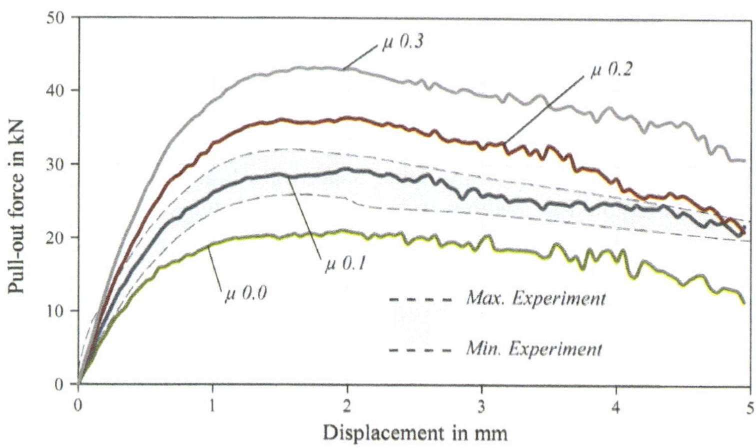

Fig. 9 Force–displacement curves of the hybrid joints with macro form closure for different friction coefficients in the interface

3.2.2 Form Closure on the Mesoscale

Similar to the macro contour, two different form elements (compare Fig. 4) are investigated on the mesoscale with regard to their load-bearing behaviour. For a targeted design of the meso structures, essential relationships between geometric parameters and the failure of the connection at maximum force transmission are analysed using numerical parameter studies on simplified models with only a single form closure element.

The first meso structure is a circumferential ring groove, which is described by the parameters width W and form closure height H as well as the edge radius R. The width of the groove is adapted to be compatible with the macro contour and is $W = 3.5$ mm (Fig. 10). The other geometric parameters are varied in the simulations.

The results are shown in Fig. 11 and demonstrate a significant influence of the edge radii on both the transferable forces (a) and the joint stiffness (b) as well as on the failure in the fibre composite. With small radii, high material utilisation can be achieved through the superposition of local fibre damage and plastic deformation in the aluminium. As the radii increase, the joint fails by reducing the load-bearing form closure component without complete failure in the fibre composite.

The second form closure element investigated on the mesoscale has a rhomb shape. The tip angle of $\alpha = 30°$ is adapted to the braiding angle of the tapes and the length of the rhomb $L_R = 6.9$ mm is adapted to the flanks of the macro contour. The edge radii R_R and the height of the rhomb H_R were varied while the radii of the corners R_{RL} and R_{RC} were kept constant at 0.5 mm. Figure 12 shows the scaled transmittable pull-out forces and the joint stiffness as a function of the rhomb height and the rhomb radii.

The greatest forces can be transmitted at maximum rhombus height at medium radii. The pull-out forces decrease with increasing radii due to an earlier loss of the form closure. Accordingly, the damages in the first layer of the CFRP part in fibre direction indicate that a high utilisation of the material can be realised with a small to medium radius (Fig. 13). On the other hand, for a high stiffness of the joint, a small radius is required at medium to large rhombus heights.

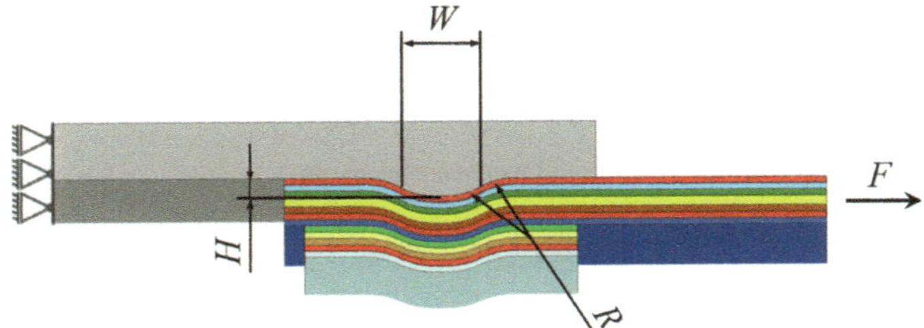

Fig. 10 Section view including the characteristic dimensions of the meso groove structure

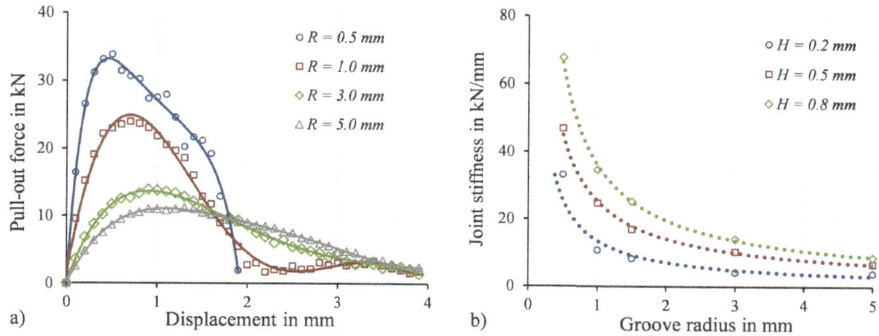

Fig. 11 **a** Force–displacement curves of a meso-groove structure for different radii with a form closure height $H = 0.8$ mm and **b** influence of the groove radius on the joint stiffness for different form closure heights

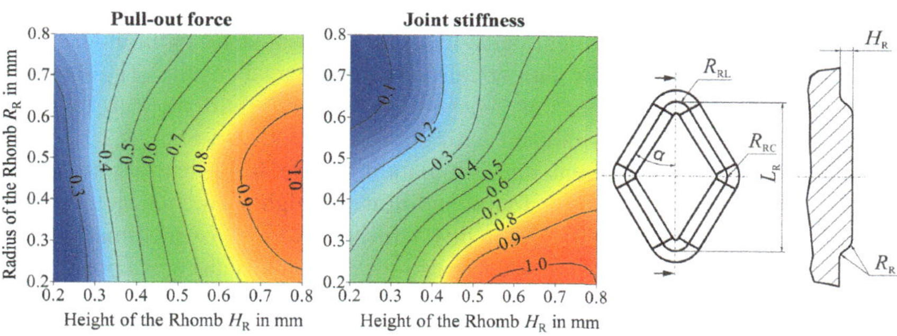

Fig. 12 Influence of rhomb radius and height on pull-out force and joint stiffness

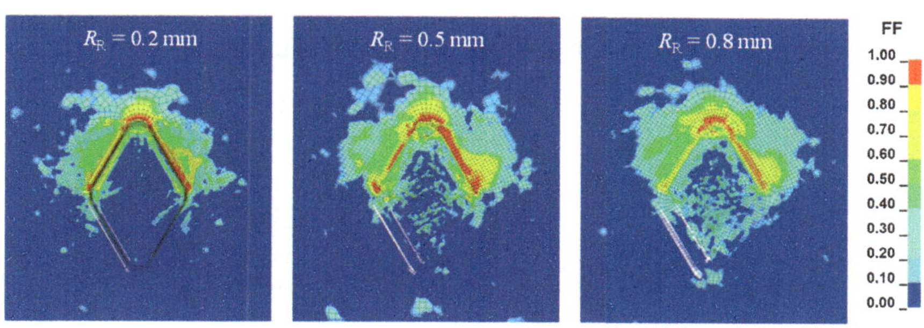

Fig. 13 Fracture criterion for fibre fracture in the first layer of the laminate for different rhomb radii at maximum force for a height of 0.8 mm

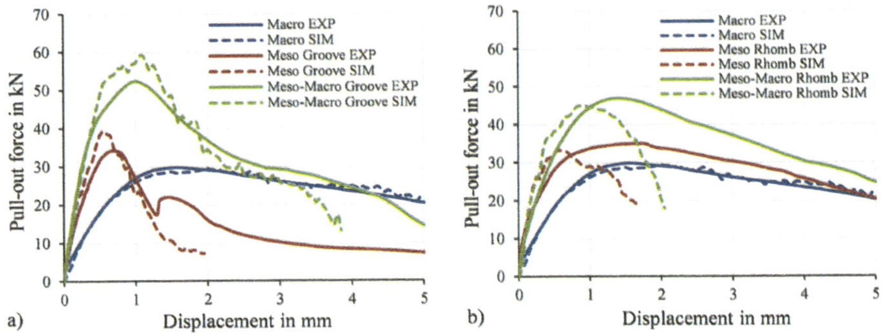

Fig. 14 Force–displacement curves from simulation and experiment for the **a** groove- and **b** rhomb meso-, macro- and combined structures

3.2.3 Combined Form Closure on Meso and Macro Scale

In Fig. 14 the force–displacement behaviour during quasi-static pull out tests of the hybrid joints is shown for the numerical and experimental results. The numerical simulations of the meso, macro and meso-macro variants is done with the coefficient of friction of $\mu = 0.1$ validated with the macro model in Sect. 3.2.1. A good agreement between experiment and simulation of the load bearing and failure behaviour during pull-out can be found and thus the high prediction quality of the friction-based modelling approach can be confirmed. A slight overestimation of the pull-out force for the groove meso and meso-macro combinations is apparent. The rhomb meso and meso-macro combinations fail earlier with a more rapid loss of pull-out force compared to the experimental results. A similar load level with higher joint stiffness compared to the macro structured joint was achieved experimentally and could be validated numerically for the meso structured joints. Compared to meso structured specimen the joint's stiffness is further increased by combining meso and macro structuring. The investigations have shown that a multi-scale structuring of LI-elements in hybrid joints enable a significant improvement in stiffness and strength.

4 Conclusion

The use of hybrid joints with multi-scale form closure enables the transmission of high loads in weight-optimised lightweight structures. In this work, the load bearing capability of manufactured hybrid joints with a tape-braided reinforced hollow profile made of CF/PA6 in combination with a multi-scale structured aluminium LI-element is analysed numerically and experimentally. A simulative approach to predict the failure of such structures is presented. With the use of an "elastic damage" model based on degradation theory a good agreement between experimental and simulative results of the load bearing capability and damage propagation in quasi-static pull out test is achieved. The tensile strength and stiffness of the different meso and macro structured form closure variants is predicted and the benefit of a multi-scale structured form closure is demonstrated. A slight overestimation of the pull-out force for the meso and meso-macro variants as well as a premature failure behaviour compared to the experimental results indicates a small inaccuracy of the numerical model. In numerical sensitivity analyses with varying height and radii of the groove and rhomb meso structures it is found that the joint strength and stiffness can be increased if the radii are decreased and the structure height is increased.

Future investigations should focus on feasibility studies of the found optimal geometric dimensions of the meso structures in terms of formability of the aluminium during hydroforming. Additionally, numerical and experimental studies of the CFRP in the bladder-assisted moulding process can give insights on the moulding behaviour and whether these optimised geometries show a good moulding quality of the meso structures, which is necessary to achieve a high load bearing capability of the contour joint. Level set tests in combination with photomicrographs and computed tomography should be carried out to generate a deeper knowledge for the phenomena during forming. Additional effort should be contributed towards the numerical rhomb model in order to increase the accuracy of the prediction of the failure behaviour. The modelling of the braided textile and the undulation of the braided plies are subjects of future investigation.

Acknowledgements The authors gratefully acknowledge the funding of this work by the German Research Foundation, DFG, within the priority programme 1712: Intrinsic hybrid composites for lightweight structures—basics of manufacturing, characterization and design.

References

Baburaj, E.G., Starikov, D., Evans, J., Shafeev, G.A., Bensaoula, A.: Enhancement of adhesive joint strength by laser surface modification. Int. J. Adhes. Adhes. **27**(4), 268–276 (2007). https://doi.org/10.1016/j.ijadhadh.2006.05.004

Barfuss, D., Garthaus, C., Gude, M., Grützner, R.: Design of multi-scale-structured Al-CF/PA6 contour joints. IJAUTOC **2**(3/4), 299–315 (2016). https://doi.org/10.1504/IJAUTOC.2016.10005317

Barfuss, D., Würfel, V., Grützner, R., Gude, M., Müller, R.: Integral blow moulding for cycle time reduction of CFR-TP aluminium contour joint processing. In: Fratini, L., Di Lorenzo, R., Buffa, G., Ingarao, G. (eds.) Proceedings of the 21st International ESAFORM Conference on Material Forming. ESAFORM 2018, Palermo, 23–25 April 2018. AIP Conference Proceedings. AIP Publishing, Melville, New York (2018). https://doi.org/10.1063/1.5034876

Fleischer, J., Koch, S.-F., Coutandin, S.: Manufacturing of polygon fiber reinforced plastic profiles by rotational molding and intrinsic hybridization. Prod Eng. Res. Dev. **9**(3), 317–328 (2015). https://doi.org/10.1007/s11740-015-0620-0

Fleischer, J., Coutandin, S., Nieschlag, J.: Einführung in intrinsische Hybridverbunde. In: Fleischer, J. (ed.) Intrinsische Hybridverbunde für Leichtbautragstrukturen, pp. 1–9. Springer Vieweg, Berlin, Heidelberg (2021)

Garthaus, C., Barfuss, D., Witschel, B., Gude, M.: Tape braiding: high-performance fibre-reinforced thermoplastic profile structures. JEC Compos. Mag. **96**, 62–64 (2015)

Gebhardt, J., Fleischer, J.: Experimental investigation and performance enhancement of inserts in composite parts. Procedia CIRP **23**, 7–12 (2014). https://doi.org/10.1016/j.procir.2014.10.084

Grützner, R., Würfel, V., Müller, R., Gude, M.: Combined external and internal hydroforming process for aluminium load introduction elements in intrinsic hybrid CFRP contour joints. In: Dröder, K., Vietor, T. (eds.) Technologies for Economic and Functional Lightweight Design. Zukunftstechnologien für den multifunktionalen Leichtbau, pp. 253–266. Springer Vieweg, Berlin, Heidelberg (2021)

Gude, M., Lenz, F., Gruhl, A., Witschel, B., Ulbricht, A., Hufenbach, W.: Design and automated manufacturing of profiled composite driveshafts. Sci. Eng. Compos. Mater. **22**(2), 187–197 (2014). https://doi.org/10.1515/secm-2014-0048

Gude, M., Barfuß, D., Coutandin, S., Fleischer, J., Grützner, R., Hirsch, F., Kästner, M., Müller-Pabel, M., Müller, R., Nieschlag, J., Ruhland, P., Würfel, V.: Hybride Hohlstrukturen für Wellen und Streben. In: Fleischer, J. (ed.) Intrinsische Hybridverbunde für Leichtbautragstrukturen, pp. 205–264. Springer Vieweg, Berlin, Heidelberg (2021)

Hallquist, J.: LS-DYNA Theory Manual: Livermore Software Technology Corporation (LSTC), California (2019)

Hashin, Z.: Failure criteria for unidirectional fiber composites. J. Appl. Mech. **47**(2), 329–334 (1980). https://doi.org/10.1115/1.3153664

Herwig, A., Schmidt, C., Horst, P.: Investigation of asymmetrical fiber metal hybrids used as load introduction element for thin-walled CFRP structures. Procedia CIRP **99**, 443–447 (2021). https://doi.org/10.1016/j.procir.2021.03.063

Hirsch, F., Kästner, M.: Microscale simulation of adhesive and cohesive failure in rough interfaces. Eng. Fract. Mech. **178**, 416–432 (2017). https://doi.org/10.1016/j.engfracmech.2017.02.026

Hufenbach, W., Kroll, L., Gude, M.: Integrated load introduction system for composite lightweight structures in aerospace applications. In: 27th SAMPE Europe International Conference, Paris (2006)

Hufenbach, W., Gude, M., Helms, O., Ulbricht, A.: Novel lightweight aircraft components in multi-material design with adapted load-introduction systems. In: European Conference on Materials and Structures in Aerospace (EUCOMAS), Berlin (2008)

Jackson, K.E., Littell, J.D., Fasanella, E.L.: Simulating the impact response of composite airframe components. In: Proceedings of the 13th LS-DYNA International Conference, Detroit (2014)

Kießling, R., Ihlemann, J., Riemer, M., Drossel, W.-G., Scharf, I., Lampke, T., Sharafiev, S., Pouya, M., Wagner, M.F.-X.: The interface of an intrinsic hybrid composite—development, production and characterisation. Procedia CIRP **66**, 289–293 (2017). https://doi.org/10.1016/j.procir.2017.03.288

Livermore Software Technology Corporation: LS-Dyna Keyword—Users Manual Volume II: Material Models. LS-DYNA R11 (2019)

Matheis, R.: Beitrag zum Fortschritt im Automobilleichtbau durch die Entwicklung von Crashabsorbern aus textilverstärkten Kunststoffen auf Basis geflochtener Preforms und deren Abbildung in der Simulation. FAT-Schriftenreihe, vol. 246. Technische Informationsbibliothek u. Universitätsbibliothek; VDA, Hannover, Berlin (2013)

Matzenmiller, A., Lubliner, J., Taylor, R.L.: A constitutive model for anisotropic damage in fiber-composites. Mech. Mater. **20**(2), 125–152 (1995). https://doi.org/10.1016/0167-6636(94)00053-0

Nieschlag, J., Ruhland, P., Daubner, S., Koch, S.-F., Fleischer, J.: Finite element optimisation for rotational moulding with a core to manufacture intrinsic hybrid FRP metal pipes. Prod. Eng. Res. Dev. **12**(2), 239–247 (2018). https://doi.org/10.1007/s11740-017-0788-6

Nieschlag, J., Eisenhardt, P., Coutandin, S., Fleischer, J.: Numerical design of rotationally molded composite tie rods. Compos. Struct. **278**,114687 (2021). https://doi.org/10.1016/j.compstruct.2021.114687

Oden, J.T., Martins, J.A.C.: Models and computational methods for dynamic friction phenomena. Comput. Methods Appl. Mech. Eng. **52**(1–3), 527–634 (1985). https://doi.org/10.1016/0045-7825(85)90009-X

Schürmann, H.: Konstruieren mit Faserverbundwerkstoffen, 1st edn. Springer, Berlin Heidelberg, Berlin, Heidelberg (2007)

Schweizerhof, K., Weimar, K., Munz, Th., Rottner, Th.: Crashworthiness analysis with enhanced composite material models in LS-DYNA—merit and limits. In: Proceedings of the 5th LS-DYNA International Conference, Southfield, Michigan (1998)

Wang, Z., Bobbert, M., Dammann, C., Zinn, C., Lauter, C., Mahnken, R., Meschut, G., Schaper, M., Troester, T.: Influences of interface and surface pretreatment on the mechanical properties of metal-CFRP hybrid structures manufactured by resin transfer moulding. IJAUTOC **2**(3/4), 272 (2016). https://doi.org/10.1504/IJAUTOC.2016.084323

Würfel, V., Barfuss, D., Grützner, R., Hirsch, F., Gude, M., Müller, R.: Design and intrinsic processing of a hybrid CFR-TP contour joint with a multi-scale structured load introduction element. In: Hausmann, J.M., Siebert, M., von Hehl, A. (eds.) Hybrid—Materials and Structures 2018—Proceedings. Hybrid 2018, Bremen, 18–19 April 2018, pp. 262–268 (2018)

Würfel, V., Grützner, R., Hirsch, F., Barfuss, D., Gude, M., Müller, R., Kästner, M.: Hybrid fibre reinforced thermoplastic hollow structures with a multi-scale structured metal load introduction element. In: Hausmann, J.M., Siebert, M., Hehl, A. von, Weidenmann, K.A. (eds) Hybrid—Materials and Structures 2020—Proceedings. Hybrid 2020, 28–29 April 2020, pp. 138–143 (2020)

Xiao, X.: Modeling energy absorption with a damage mechanics based composite material model. J. Compos. Mater. **43**(5), 427–444 (2009). https://doi.org/10.1177/0021998308097686

Xiao, X., Botkin, M.E., Johnson, N.L.: Axial crush simulation of braided carbon tubes using MAT58 in LS-DYNA. Thin-Walled Struct. **47**(6–7), 740–749 (2009). https://doi.org/10.1016/j.tws.2008.12.004

Zinn, C., Bobbert, M., Dammann, C., Wang, Z., Tröster, T., Mahnken, R., Meschut, G., Schaper, M.: Shear strength and failure behaviour of laser nano-structured and conventionally pre-treated interfaces in intrinsically manufactured CFRP-steel hybrids. Compos. B Eng. **151**, 173–185 (2018). https://doi.org/10.1016/j.compositesb.2018.05.030

Model-Based Joining Process Design for the Body Shop Process Chain

Christian Schwarz, Patrick Ackert, Tobias Falk, Markus Puschmann, Reinhard Mauermann, and Welf-Guntram Drossel

Abstract The future mobility presents new safety requirements. In addition to the passenger compartment, batteries and hydrogen tanks need also to be intensively protected. High-strength, hot-formed steels, die-cast components and new material pairings pose new challenges for selecting and parametrization the used joining technology. For this reason, the Coupled Process Analysis (CPA) method, for supporting the design and monitoring of joining processes, is presented. From as early as 12 data sets from simulation or experiment (micrograph), it is possible to map interdependencies between the quality criteria of the join and the acting process parameters. The resulting possibilities are illustrated by means of the numerical design of a clinch and a riveted joint as well as the experimental, image-based sampling of a weld seam. Finally a supplemented by a presentation of the potentials in quality monitoring in series operation of the automotive process chain will be done. The combination of high flexibility with respect to the in- and output variables (visual images, FE-meshes, process curves, discrete values), low modeling effort and the image based representation of the interdependencies makes the approach suitable even for employees with lower qualification.

Keywords Joining · Car body shop process chain · Surrogate modelling · Quality optimization

1 Introduction

The reduction of development cycles is an essential criterion for achieving competitive advantages in automotive engineering. At the same time, the growing use of

C. Schwarz (✉) · P. Ackert · T. Falk · M. Puschmann · R. Mauermann · W.-G. Drossel
Fraunhofer Institute for Machine Tools and Forming Technology IWU, Reichenhainer Str. 88, 09126 Chemnitz, Germany
e-mail: christian.schwarz@iwu.fraunhofer.de

W.-G. Drossel
Institute for Machine Tools and Production Processes, TU Chemnitz, Reichenhainer Str. 70, 09126 Chemnitz, Germany

L. F. M. da Silva et al. (eds.), *2nd International Conference on Advanced Joining Processes (AJP 2021)*, Proceedings in Engineering Mechanics,
https://doi.org/10.1007/978-3-030-95463-5_2

mixed construction concepts in body manufacturing (e.g. steel/aluminum sheet) is increasing complexity in production, which significantly raises the requirements for meeting quality targets.

Digitization in production is growing steadily. The development process generates data from experimental tests and numerical simulations. This data is used to determine suitable process parameters for manufacturing methods such as joining technology. More and more process, measurement and simulation data is available to manufacturers, but there are still methods missing to gain useful insights. Especially when the test sample size is small, the process data is incomplete and the type of recorded data differs, there is no satisfactory solution approach available. The best example of this is a glance at vehicle production where up to 500 stamped sheet metal parts are joined in up to 120 fixtures to get a complete body, and each process step has an varying level of impact on the dimensional stability of the car body (Hu et al. 2001).

2 Statistical Modelling in Joining

2.1 Mechanical Joining

Machine learning methods have already been used in research work on mechanical joining technology. Oudjene presents a combination of response surface method and moving least square approximation to optimize the tool contour during clinching (Oudjene and Ben-Ayed 2008; Oudjene et al. 2009). In a sensitivity analysis performed by Lambiase, the main effects of tool parameters on undercut and neck thickness are determined by an artificial neural network (ANN) (Lambiase and Di Ilio 2013). Also, based on a genetic algorithm (GA), the tool geometry is optimized with respect to the joint strength. Roux optimizes the tool geometry with respect to strength parameters of the joining partners using a kriging-based model (Roux and Bouchard 2013). For the targeted design of clinching processes, Drossel investigate the application of an FE-based approach for sensitivity analysis (Drossel et al. 2014). The objective of the study was to optimize the strength-related quality parameters for different sheet thickness and material combinations. Wang uses Bezier curves parameterized in their control points to describe the die geometry. By varying the control points, die geometry can be optimized to avoid crack formation when clinching thin sheets (Wang et al 2017). Schwarz propose an optimization approach based on principal component analysis to improve the quality of the joined joint (Schwarz et al. 2020). In Thoms and Kalich (2002), neural networks were used to predict process parameters as well as joint strengths in solid stamp riveting. In the studies, Hahn and Tan (2003), Hahn et al. (2005) and Tan et al. (2006), neural networks and later k-nearest neighbor regressions were first used to predict the load carrying capacity of clinch joints. In Breckweg (2006) neural networks to classify defects in process monitoring for radial clinching are used. For self-pierce-riveting, the influence of fluctuating component properties, such as sheet thickness and strength variations, on the joining point formation was investigated and concepts for adaptive process adjustment were demonstrated (Jäckel et al. 2016).

2.2 *Welding*

Three main methods are used for statistical modeling of experimental welding data: Response Surface Method, Taguchi Method and Artificial Neural Networks (Dhas and Dhas 2012).

A popular method for the mathematical description of cause-effect relationships in welding is the response surface method (RSM). Here, a multi-dimensional response surface is used to establish a relationship between the input parameters and one or more quality-relevant target variables. To plan the experiments, a full factorial experimental design is often used in RSM with the Central Composite Design (CCD). The first experimental studies based on RSM are shown by Gunaraj and Murugan (1999) to identify the optimum weld geometry in submerged arc welding. The welding parameters considered within the models are clamping voltage, wire feed rate, welding speed and welding nozzle distance to the sheet.

Xu use RSM for weld optimization of thick-walled pipes joined by gas shielded arc welding. The objective is to establish a relationship between six welding parameters and three weld parameters determined in the micrograph (Xu et al. 2014). In Liao et al. (2018), RSM is used to predict the ideal weld geometry in TIG welding of pipelines. The geometry of the weld is analyzed using three parameters (weld depth, weld width, weld protrusion) in the micrograph. Welding power, welding speed and welding position are identified as important input variables for the model. In contrast, the optimization of the laser beam welding process is based on the RSM of Padmanaban and Balasubramanian (2010). The focus is on the model-based prediction of the tensile strength of a laser beam welded magnesium alloy with variation of laser power, welding speed and focus position. Further experimental investigations based on the RSM can be found, for example, in Acherjee et al. (2009) and Srivastava and Garg (2017).

Another statistical method that is frequently used in welding parameter studies is the Taguchi Method (TM). The TM is an experimental design that aims at minimizing the experimental scatter as well as the robust design of the process (Dhas and Dhas 2012). Rao use the TM in combination with a multiple regression model for arc welding. Input variables for the model are wire feed rate, workpiece thickness, pulse current intensity, and welding speed. Experimental results show that the model can predict the quality relevant parameters of the weld geometry with a high accuracy (Rao et al. 2009). In Anawa and Olabi (2008) the focus is on the optimization of laser beam welded components. The influence of the laser power, the laser speed and the focus position on the weld geometry is investigated. The functional relationship between welding parameter and weld is realized via nonlinear regression models. In contrast, using laser welded aluminum alloys as an example, Pan show that TM can be used to increase the tensile strength of the welded joint (Pan et al. 2005). Furthermore, Eşme (2009) describes the applicability of TM to optimize the tensile shear strength for spot welded geometries. Datta attempt to combine TM with PCA to convert the multivariate optimization problem into a univariate one for submerged arc welding (Datta et al. 2009). On the other hand, a comparison between the RSM and the TM is shown in Sivaraos Milkey et al. (2014).

In addition to RSM and TM, numerous studies deal with the optimization of welding parameters using ANN. Mirapeix shows in Mirapeix et al. (2007) a diagnostic method of welding defects based on a KNN. The quality of the weld is evaluated using the plasma spectrum. Lightfoot develop a KNN architecture that can be used to reduce the weld-induced distortion of marine steel plates (Lightfoot et al. 2004). In Ismail et al. (2013) a KNN-based model is presented which is able to predict the weld geometry characterized by two specific parameters for laser beam microwelding of thin sheet structures. Acherjee determine the correlation between process parameters and quality related target variables based on a nonlinear KNN model for laser transmission welding of thermoplastic (Acherjee et al. 2011). Laser power, laser speed, working distance and clamping pressure are selected as welding parameters for the model. Model target variables considered are tensile shear strength and weld width. A very comprehensive review on statistical analysis of welding processes can be found in Benyounis and Olabi (2008) and Dhas and Dhas (2012).

Due to the possibility of numerical calculation of welding processes, optimization of welding process parameters can also be performed based on computer simulations. In this context, the potential of welding simulation lies in a reduction of time-consuming and cost-intensive welding experiments (Yang et al. 2018). Against this background, Yang et al. (2018) and Jiang et al. (2016) propose an FEA-based method for model-based prediction of weld geometry for laser beam welding. Kriging models form the basis of the statistical investigations, with laser power, laser speed, and focus position selected as input parameters in both cases. On the other hand, an approach to combine FEA and RSM to optimize the temperature field and weld geometry for laser beam welding is shown in Acherjee et al. (2012).

2.3 *Data-Based Forecasting Models in Automotive Car Body Production*

In the context of digitalization, the possibilities of data-based analysis and prognosis in car body production are increasing. In previous development projects on the digitization of car body production, for example, the influence of fluctuating sheet metal properties on component quality and geometry was investigated in deep drawing (Purr et al. 2015; Kriechenbauer et al. 2020). It was shown that an adaptive reaction to process fluctuations in deep drawing can be realized with actuators integrated in the tools (Bäume et al. 2016). Schwarz describe that it is possible to predict the interactions of process parameters on the component geometry for deep drawing using data-based statistical models (Schwarz et al. 2018, 2021).

3 Coupled Process Analysis

Traditionally, quality assurance is the instance in which most of the data for the final product is determined. The classic implementation is mostly carried out on a random sample basis and under laboratory conditions. In the course of Industry 4.0, more and more data is now being collected along the production line. This data includes product data (dimension, strength, etc.) from the real process (Behrens et al. 2018), as well as simulation data, which is collected in an early planning phase (Wahl et al. 2011). The aim of the data acquisition is to determine the effect of process parameter variations on quality-relevant target variables (e.g. dimensional accuracy, strength values).

However, the statistical analysis of these data is very challenging in terms of data diversity and data volume. In most cases, the results are in the form of multidimensional data sets (e.g. finite element (FE) meshes from simulations). If the simulation result (e.g. deviation, temperature, stress) is determined at each mesh node of the FE mesh, the connection of the process parameters with this target variable has to be considered as a multivariate statistical problem due to the large number of nodes in a FE mesh.

To solve this problem, a machine learning method already exists in the form of Coupled Process Analysis (CPA), which deals with the evaluation of multidimensional data sets along production process chains. The aim of the CPA modeling algorithm is to combine the dependent and independent data sets along any multi-stage process chain so that the causal relationships can be determined using surrogate models.

The solution consists in projecting the multivariate data basis (e.g. sheet thickness distribution, deviation field) into a low-dimensional subspace by means of Fourier- and variance-based reduction methods as shown in Schwarz et al. (2018, 2020, 2021) (see Fig. 1). In this subspace, computation of surrogate models and sensitivities is much less complex due to the low dimension. The statistical quantities calculated in the subspace can then be visualized as a shape-based area plot via a back-projection process onto the reference mesh. This allows the functional relationship between the process parameters and the defined target variable to be visualized in a spatially resolved manner for each node of the finite element mesh.

The strength and focus of CPA lie in the creation of surrogate models by applying efficient data reduction methods. In summary, the application of the CPA results in the following advantages:

- Data diversity: Processing of multivariate input data (discrete values, mesh surface data, images).

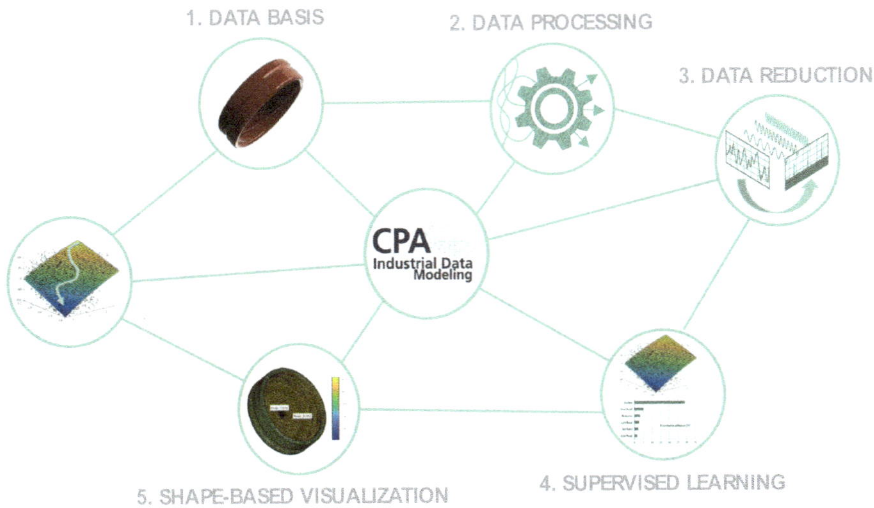

Fig. 1 Methodology of coupled process analysis (CPA)

- Data size: The CPA requires only a very small sample size for a large number of input variables.
- Range of applications: The CPA can be used both in the planning stage and in series production.
- Shape-based visualization: User-friendly representation of complex inter-relationships.

The possibility to process and evaluate multivariate data sets results in a higher gain of knowledge, whereby the set quality targets can be reached faster.

4 Application of CPA Algorithm in Joining

4.1 Laser Beam Welding

The objective of the first case study is the development of a process-specific software demonstrator for the virtual prediction of both welding parameters to be set (given welding results) and multivariate welding results (given parameters) based on surrogate models for the example of laser beam welding. This goal is to be achieved by developing an approach based on CPA with which the complete two-dimensional weld geometry can be processed as a multivariate data set and statistically evaluated.

In contrast to the state of the art, the welding parameters are to be linked to multivariate welding result variables with the aid of selected model functions, which can consist of over one million dependent result values (see Fig. 2). To verify the func-

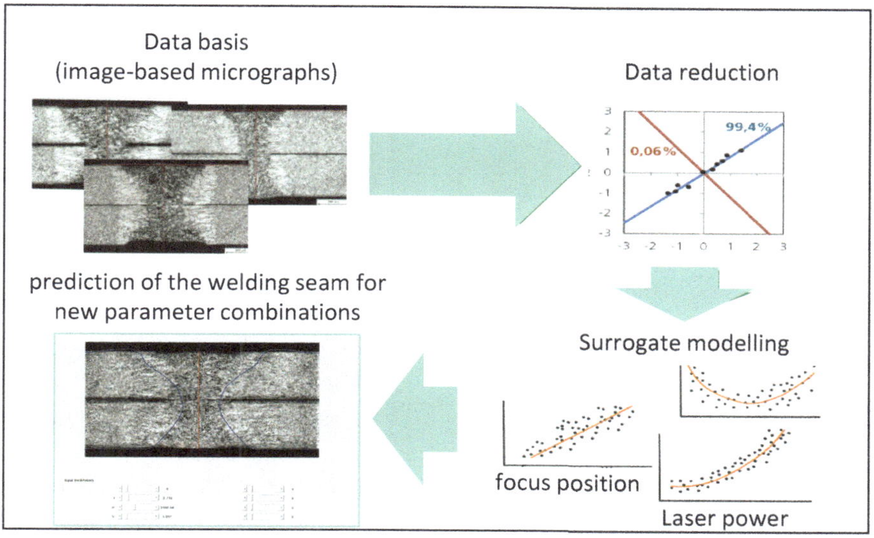

Fig. 2 Model-based prediction of micrographs using the CPA algorithm

tionality of the CPA, three welding process parameters (laser power, focus position, laser speed) were varied in the experiment and only 12 real image-based micrographs were generated as data basis.

It has been demonstrated that it is possible to process entire image micrographs or even weld cross-section contours by statistical models and thus to predict complex welding results, such as the geometry of the heat-affected zone or/and the fusion area, and to visualize them in a 2D result plot.

4.2 *Clinching*

FEM-based parameter studies have great relevance in the design of clinch tools. Often, only a small number of target parameters such as geometry parameters, strength values or forces are taken into account to analyze the quality of the joined joint. On the basis of CPA, a description of the entire clinch geometry becomes possible via a mathematical model. The aim of the optimization is to maximize the interlock and neck thickness for the entire clinch contour.

For the process design of the clinching process, a 2D simulation model was created using the FE solver Deform. The FE model is parameterized in such a way that the input parameters can be systematically varied within defined parameter limits during each simulation run. As input parameters, ten tool and process parameters (e.g. diameter of the die, diameter of the punch, blank thickness), were varied. To

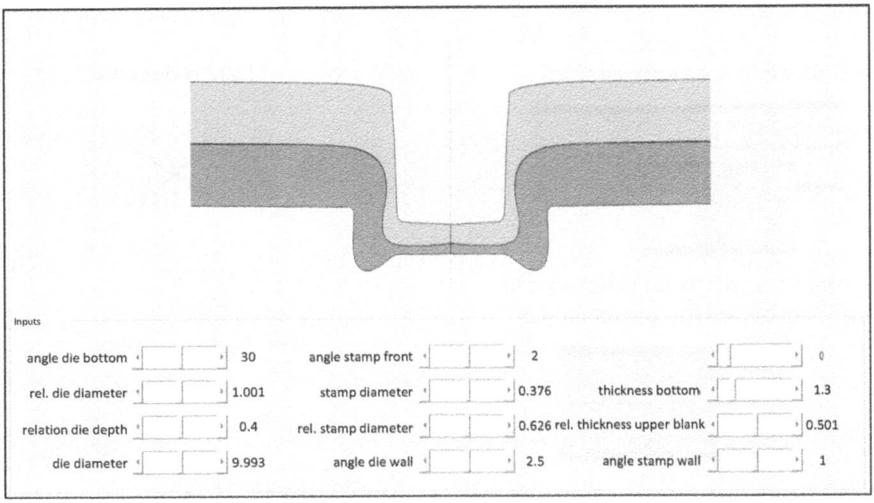

Fig. 3 Model-based prediction of the 2D cross-sectional geometry of the clinched joint

achieve the most uniform distribution of samples in the experimental space, a sample set with 100 design variants was generated using Latin Hypercube Sampling (LHS).

As shown in Fig. 3, the CPA algorithm is able to create a surrogate model that mathematically generates the 2D cross-sectional geometry of the mechanical joint in real time as the tool parameters change freely in parameter space. The interactions between tool parameters and clinch contour can be reproduced with a very high forecasting quality.

Furthermore, it could be shown that the additional application of an evolutionary algorithm offers the possibility to optimize the entire clinch contour with regard to relevant joint parameters such as undercut and neck thickness.

4.3 Self Pierce Riveting

The aim of the research project was to develop a FE-solver-independent joining assistance system for self-pierce riveting that visualizes the functional relationship between process parameters and user-defined result variables such as joining point geometry, crack formation or tool load as a 2D cross-sectional model in real time on the basis of joining process simulation data.

To build the database, a rotationally symmetric 2D simulation model was constructed using the software Simufact Forming and 150 simulation variants were calculated. A total of 14 process parameters (e.g. blank thickness, diameter die, riveting height) were modified in the simulation. For a user-friendly use of the CPA models, a software demonstrator was built (ref. Fig. 4). In this demonstrator, the 14

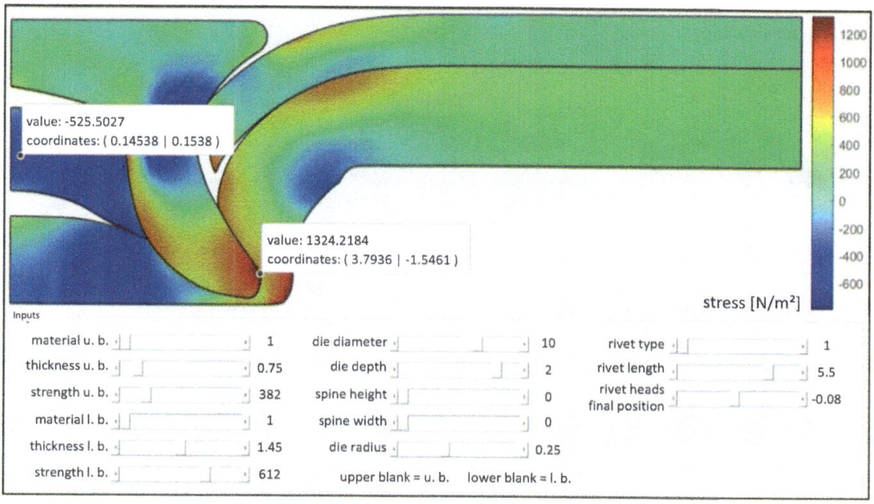

Fig. 4 Model-based prediction of the 2D cross-sectional geometry of a self-pierce riveted joint

input parameters are set using sliders and, based on this input data set, the predicted joining point contour is immediately visualized. Based on this geometry, the characteristic parameters and the maximum joining force of the predicted joint are automatically calculated. In addition, the user can have the predictions of all integrated output variables displayed. These are visualized node/element-wise via color gradients, as one is used to from commercial simulation software. In addition, it is possible to directly display the value of the currently selected output size for nodes in the software by clicking on the corresponding accounts. If another output size is selected, then not only the coloring/scale will change, but also the value of the selected node will be updated.

4.4 Car Body Manufacturing

In this case study CPA algorithm is used to identify causal relationships in industrial process chains such as car body manufacturing. The database comprises inline measurement data and fixture adjustments for a series system in which approx. 600 cars (Porsche Panamera) were built. The objective was to determine the causal effects of the measured values and the fixture adjustments of the steps "front section", "rear section", "floor center" and the assembly "underbody" on the "body assembly" (Fig. 5). The "side panels" are measured separately and included in the process input.

It is shown that by applying the CPA method to a car body process chain, it was possible to combine dependent data sets (e.g. geometry) and independent input data sets (e.g. fixture adjustments) to form a surrogate model. The cause of variations in

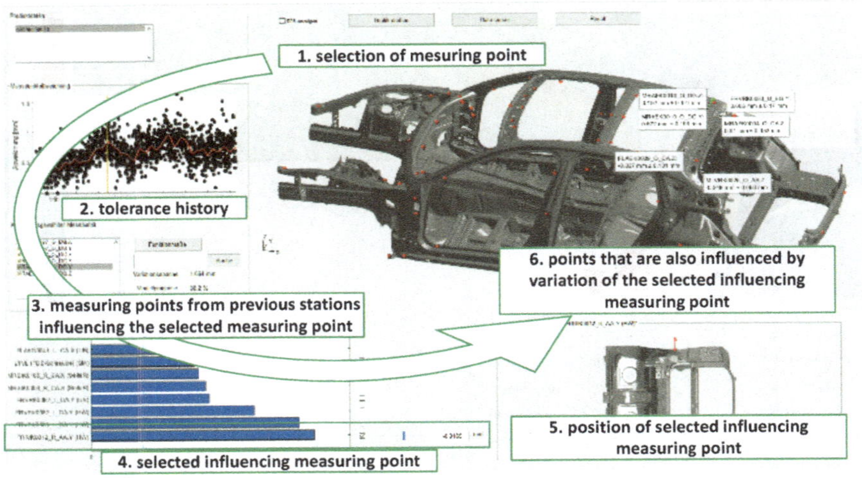

Fig. 5 Applying CPA in car body manufacturing process

the assembly stage could be identified and used to derive compensation measures. The accuracy of the final car body assembly and the process chain efficiency are significantly improved.

5 Conclusion

Digitization is steadily growing in joining processes. More and more data on the process, measurement and simulation is available to manufacturers. The aim of the data acquisition is to determine the effect of process parameter variations on quality-relevant target variables (e.g. dimensional accuracy, strength values).

Within the paper, Coupled Process Analysis (CPA) was presented for the statistical analysis of joining processes. The CPA is a machine learning method that is designed to recognize quality-relevant process parameters in a few simulation or production runs and to make corresponding compensation suggestions. The strength and focus of CPA lie in the creation of surrogate models by applying efficient data reduction methods. This allows multivariate data structures (images, in-line measurement data, FE-meshes) to be processed and complex relationships to be visualized with maximum information content for the user.

Moreover, a verification shall be conducted regarding the ability of CPA to determine correlations in other joining processes. Furthermore, the transferability of the CPA method to other industrial processes can be examined, e.g. in medicine, the chemical.

References

Acherjee, B., Misra, D., Bose, D., Venkadeshwaran, K.: Prediction of weld strength and seam width for laser transmission welding of thermoplastic using response surface methodology. Opt. Laser Technol. **41**(8), 956–967 (2009). https://doi.org/10.1016/j.optlastec.2009.04.007

Acherjee, B., Mondal, S., Tudu, B., Misra, D.: Application of artificial neural network for predicting weld quality in laser transmission welding of thermoplastics. Appl. Soft Comput. **11**(2), 2548–2555 (2011). https://doi.org/10.1016/j.asoc.2010.10.005

Acherjee, B., Kuar, A.S., Mitra, S., Misra, D.: Modeling and analysis of simultaneous laser transmission welding of polycarbonates using an FEM and RSM combined approach. Opt. Laser Technol. **44**(4), 995–1006 (2012). https://doi.org/10.1016/j.optlastec.2011.10.018

Anawa, E.M., Olabi, A.G.: Using Taguchi method to optimize welding pool of dissimilar laser-welded components. Opt. Laser Technol. **40**(2), 379–388 (2008). https://doi.org/10.1016/j.optlastec.2007.07.001

Bäume, T., Zorn, W., Drossel, W.-G., Rupp, G.: Iterative process control and sensor evaluation for deep drawing tools with integrated piezoelectric actuators. Manuf. Rev. **3**(H3) (2016). https://doi.org/10.1051/mfreview/2016002

Behrens, B.-A., Groche, P., Krüger, J., Wulfsberg, J.P.: WGP-Standpunkt Industriearbeitsplatz 2025. Technical Report, WGP Wissenschaftliche Gesellschaft für Produktionstechnik e.V., Hannover (2018)

Benyounis, K.Y., Olabi, A.G.: Optimization of different welding processes using statistical and numerical approaches—a reference guide. Adv. Eng. Softw. **39**(6), 483–496 (2008). https://doi.org/10.1016/j.advengsoft.2007.03.012

Breckweg, A.: Automatisiertes und prozessüberwachtes Radialclinchen höherfester Blechwerkstoffe. Doctoral thesis, University of Stuttgart (2006)

Datta, S., Nandi, G., Bandyopadhyay, A., Kumar Pal, P.: Application of PCA-based hybrid Taguchi method for correlated multicriteria optimization of sub-merged arc weld. A case study. Int. J. Adv. Manuf. Technol. **45**(3–4), 276–286 (2009). https://doi.org/10.1007/s00170-009-1976-0

Dhas, J.E.R., Dhas, S.J.H.: A review on optimization of welding process. Procedia Eng. **38**, 544–554 (2012). https://doi.org/10.1016/j.proeng.2012.06.068

Drossel, W.-G., Falk, T., Israel, M., Jesche, F.: Unerring planning of clinching processes through the use of mathematical methods. KEM **611–612**, 1437–1444 (2014). https://doi.org/10.4028/www.scientific.net/KEM.611-612.1437

Eşme, U.: Application of Taguchi method for the optimization of resistance spot welding process. Arabian J. Sci. Eng. **34**(2B) (2009)

Gunaraj, V., Murugan, N.: Application of response surface methodology for predicting weld bead quality in submerged arc welding of pipes. J. Mater. Process. Technol. **88**(1–3), 266–275 (1999). https://doi.org/10.1016/S0924-0136(98)00405-1

Hahn, O., Tan, Y.: Vorhersage des Tragverhaltens von Clinchverbindungen unter quasi-statischer Scherzugbelastung mittels eines neuronalen Netzes. Schweißen Und Schneiden **4**, 138–143 (2003)

Hahn, O., Tan, Y., Voight, H.-M.: Maschinelles Lernen zur Vorhersage der Tragfähigkeit von Clinchverbindungen. UTF Sci. **1** (2005)

Hu, M., Lin, Z., Lai, X., Ni, J.: Simulation and analysis of assembly processes considering compliant, non-ideal parts and tooling variations. Int. J. Mach. Tools Manuf. **41**(15), 2233–2243 (2001). https://doi.org/10.1016/S0890-6955(01)00044-X

Ismail, M.I.S., Okamoto, Y., Okada, A.: Neural network modeling for prediction of weld bead geometry in laser microwelding. Adv. Opt. Technol. **5**, 1–7 (2013). https://doi.org/10.1155/2013/415837

Jäckel, M., Falk, T., Landgrebe, D.: Concept for further development of self-pierce riveting by using cyber physical systems. Procedia CIRP **44**, 293–297 (2016). https://doi.org/10.1016/j.procir.2016.02.073

Jiang, P., Cao, L., Zhou, Q., Gao, Z., Rong, Y., Shao, X.: Optimization of welding process parameters by combining Kriging surrogate with particle swarm optimization algorithm. Int. J. Adv. Manuf. Technol. **86**(9–12), 2473–2483 (2016). https://doi.org/10.1007/s00170-016-8382-1

Kriechenbauer, S., Müller, P., Mauermann, R., Drossel, W.-G.: Extension of process limits with bidirectional deep drawing: In: Behrens, B.A., Brosius, A., Hintze, W., Ihlenfeldt, S., Wulfsberg, J.P. (eds.) Production at the Leading Edge of Technology. WGP 2020. Lecture Notes in Production Engineering. Springer, Berlin, Heidelberg (2020). https://doi.org/10.1007/978-3-662-62138-7_10

Lambiase, F., Di Ilio, A.: Optimization of the clinching tools by means of integrated FE modeling and artificial intelligence techniques. Procedia CIRP **12**, 163–168 (2013). https://doi.org/10.1016/j.procir.2013.09.029

Liao, B., Shi, Y., Cui, Y., Cui, S., Jiang, Z., Yi, Y.: Mathematical model for prediction and optimization of weld bead geometry in all-position automatic welding of Pipes. Metals **8**(10), 756 (2018). https://doi.org/10.3390/met8100756

Lightfoot, M.P., Bruce, G.J., McPherson, N.A., Woods, K.: The application of artificial neural networks to weld-induced deformation in ship plate. Welding J. 23–30 (2004)

Mirapeix, J., García-Allende, P.B., Cobo, A., Conde, O.M., López-Higuera, J.M.: Real-time arc-welding defect detection and classification with principal component analysis and artificial neural networks. NDT & E Int. **40**(4), 315–323 (2007). https://doi.org/10.1016/j.ndteint.2006.12.001

Oudjene, M., Ben-Ayed, L.: On the parametrical study of clinch joining of metallic sheets using the Taguchi method. Eng. Struct. **30**(6), 1782–1788 (2008). https://doi.org/10.1016/j.engstruct.2007.10.017

Oudjene, M., Ben-Ayed, L., Delamézière, A., Batoz, J.: Shape optimization of clinching tools using the response surface methodology with moving least-square approximation. J. Mater. Process. Technol. **209**(1), 289–296 (2009). https://doi.org/10.1016/j.jmatprotec.2008.02.030

Padmanaban, G., Balasubramanian, V.: Optimization of laser beam welding process parameters to attain maximum tensile strength in AZ31B magnesium alloy. Opt. Laser Technol. **42**(8), 1253–1260 (2010). https://doi.org/10.1016/j.optlastec.2010.03.019

Pan, L.K., Wang, C.C., Hsiao, Y.C., Ho, K.C.: Optimization of Nd. YAG laser welding onto magnesium alloy via Taguchi analysis. Opt. Laser Technol. **37**(1), 33–42 (2005). https://doi.org/10.1016/j.optlastec.2004.02.007

Purr, S., Meinhardt, J., Moelzl, K., Ostermair, M., Hagenah, H., Merklein, M.: Stamping plant 4.0—data mining for investigation and prediction of quality issues in manufacturing car body parts. In: Landgrebe, D., Drossel, W.-G., Putz, M. (eds.) 5th International Conference on Accuracy in Forming Technology and 22nd Saxon Conference on Forming Technology SFU 2015, pp. 623–642. Verlag Wissenschaftliche Scripten, Tagungsband; Auerbach (2015). https://doi.org/10.5120/12106-8375

Rao, P.S., Gupta, O.P., Murty, S.S.N., Rao, A.B.K.: Effect of process para-meters and mathematical model for the prediction of bead geometry in pulsed GMA welding. Int. J. Adv. Manuf. Technol. **45**(5–6), 496–505 (2009). https://doi.org/10.1007/s00170-009-1991-1

Roux, E., Bouchard, P.: Kriging metamodel global optimization of clinching joining processes accounting for ductile damage. J. Mater. Process. Technol. **213**(7), 1038–1047 (2013). https://doi.org/10.1016/j.jmatprotec.2013.01.018

Schwarz, C., Ackert, P., Mauermann, R.: Principal component analysis and singular value decomposition used for a numerical sensitivity analysis of a complex drawn part. Int. J. Adv. Manuf. Technol. **894**, 2255–2265 (2018). https://doi.org/10.1007/s00170-017-0980-z

Schwarz, C., Kropp, T., Kraus, C., Drossel, W.-G.: Optimization of thick sheet clinching tools using principal component analysis. Int. J. Adv. Manuf. Technol. **106**, 471–479 (2020). https://doi.org/10.1007/s00170-019-04512-5

Schwarz, C., Link, P., Ihlenfeldt, S., Drossel, W.-G.: Application of Fourier-related data reduction methods in sheet metal forming. Procedia CIRP **99**, 206–265 (2021). https://doi.org/10.1016/j.procir.2021.03.038

Sivaraos Milkey, K.R., Samsudin, A.R., Dubey, A.K., Kidd, P.: Comparison between Taguchi method and response surface methodology (RSM) in modelling CO_2 laser machining. Jordan J. Mech. Indust. Eng. **8**(1), 35–42 (2014)

Srivastava, S., Garg, R.K.: Process parameter optimization of gas metal arc welding on IS:2062 mild steel using response surface methodology. J. Manuf. Process. **25**, 296–305 (2017). https://doi.org/10.1016/j.jmapro.2016.12.016

Tan, Y., Hahn, O., Du, F., Voigt, H.: Non-destructive determination of the loading capacities of clinched DC 04 joints by kNN-regression. Mater. Test. **48**(5), 211 (2006)

Thoms, V., Kalich, J.: Prozessvorhersage beim Stanznieten mit neuronalen Netzen, EFB-Forschungsbericht Nr. 179. EFB 2002, Hannover (2002)

Wahl, M., Schulz, F., Altermann, T., Eckert, A.: Die Prozesskettensimulation - Ein Beitrag zum virtuellen Karosseriebau. In: Neugebauer, Reimund, (eds) Karosserien fertigen - nachhaltig und effizient, pp. 259–269. Verlag Wissenschaftliche Scripten, Tagungsband; Zwickau (2011)

Wang, M., Xiao, G., Li, Z., Wang, J.: Shape optimization methodology of clinching tools based on Bezier curve. Int. J. Adv. Manuf. Technol. **24**(1) (2017). https://doi.org/10.1007/s00170-017-0987-5

Xu, W.H., Lin, S.B., Fan, C.L., Zhuo, X.Q., Yang, C.L.: Statistical modelling of weld bead geometry in oscillating arc narrow gap all-position GMA welding. Int. J. Adv. Manuf. Technol. **72**(9–12), 1705–1716 (2014). https://doi.org/10.1007/s00170-014-5799-2

Yang, Y., Gao, Z., Cao, L.: Identifying optimal process parameters in deep penetration laser welding by adopting Hierarchical-Kriging model. Infrared Phys. Technol. **92**, 443–453 (2018). https://doi.org/10.1016/j.infrared.2018.07.006

Welding

Investigation of the Directional Characteristics of the Emitted Airborne Sound by Friction Stir Welding for Online Process Monitoring

Michael Grätzel, Sven Other, Benedict Stoll, Maximilian Rohe, Michael Hasieber, Torsten Löhn, Jörg Hildebrand, Jean Pierre Bergmann, András Kátai, Kati Breitbarth, and Joachim Bös

Abstract Currently, destructive or non-destructive testing methods are used to verify the weld seam quality subsequent to the manufacturing process. Therefore, pre-processes such as visible or mechanical testing require additional efforts, which can lead to expensive reworking or rejection of the components. The acoustic process

M. Grätzel (✉) · S. Other · B. Stoll · M. Rohe · M. Hasieber · T. Löhn · J. Hildebrand · J. P. Bergmann
Production Technology Group, Technische Universität Ilmenau, Gustav-Kirchhoff-Platz 2, 98693 Ilmenau, Germany
e-mail: michael.graetzel@tu-ilmenau.de

S. Other
e-mail: sven.other@tu-ilmenau.de

B. Stoll
e-mail: benedict-niklas.stoll@tu-ilmenau.de

M. Rohe
e-mail: maximilian.rohe@tu-llmenau.de

M. Hasieber
e-mail: michael.hasieber@tu-ilmenau.de

T. Löhn
e-mail: torsten.loehn@tu-ilmenau.de

J. Hildebrand
e-mail: joerg.hildebrand@tu-ilmenau.de

J. P. Bergmann
e-mail: jeanpierre.bergmann@tu-ilmenau.de

A. Kátai · K. Breitbarth (✉) · J. Bös
Fraunhofer Institute for Digital Media Technology IDMT, Ehrenbergstraße 31, 98693 Ilmenau, Germany
e-mail: kati.breitbarth@idmt.fraunhofer.de

A. Kátai
e-mail: andras.katai@idmt.fraunhofer.de

J. Bös
e-mail: joachim.boes@idmt.fraunhofer.de

L. F. M. da Silva et al. (eds.), *2nd International Conference on Advanced Joining Processes (AJP 2021)*, Proceedings in Engineering Mechanics,
https://doi.org/10.1007/978-3-030-95463-5_3

characterization for Friction Stir Welding (FSW) applications permits a comparatively new approach of process monitoring to detect weld seam irregularities by the characterization of the emitted noise in the audible frequency range (airborne sound signal). In previous publications, the acoustic detection of weld seam irregularities was mostly based on structure-borne sound sensors. Although a correlation between weld defects and audio signals has been demonstrated, there are process-related deficits in the use of structure-borne sound sensors. These include a fixed installation position and limited applicability for large-scale components such as battery cases. In contrast airborne sound sensors (microphones) can be mounted directly in the area of the joining process and thus influences of component size, joining materials, and weld seam geometry can be reduced. However, the use of airborne sound sensors for FSW applications requires preparatory considerations on the sensor position towards the joining process (sidely, in front of or behind the processing tool). Therefore, in this study an approach will be presented to evaluate the directional characteristic of the airborne sound emitted by the FSW process. First, the positioning of the microphone for the various welding directions were investigated. This was done to determine a suitable microphone orientation during the process. Then, the general determination of audio signals from the FSW process will be considered and compared to the process force feedback. Further, it was demonstrated that acoustic analysis can be used for detection of weld seam irregularities such as flash formation on 5 mm AA 5754 H111 sheets. All experiments were performed with a robotized FSW setup that was modified by a self-developed acoustic measuring device.

Keywords Machine learning · Friction stir welding · Weld seam irregularities · Process forces · Acoustic emissions · Audio signal · Polar plots · Process monitoring · Real time monitoring

1 Introduction

The joining of metallic structures by FSW (Friction Stir Welding) is gaining more and more importance, particularly for components consisting of aluminum alloys or dissimilar materials such as aluminum and copper, as these are used in e-mobility concepts. The main advantage of FSW in lightweight design is the comparative low joining temperature below the liquidus temperatures of the base materials and, thus pores and hot crack formation can be prevented. The operation principle of the FSW process is based on a wear-resistant tool consisting of a shoulder and probe. The involved process stages (I-through V) of the FSW process are shown in Fig. 1.

At the beginning of the joining process the tool is set in rotation (I) and plunges by a downward motion into the workpiece (II). During the plunging procedure the axial force F_z increases and plasticizes the material due to the frictional heat. In

J. Bös
Industrial Applications of Media Technologies IMA Group, Technische Universität Ilmenau, Am Helmholtzring 1, 98693 Ilmenau, Germany

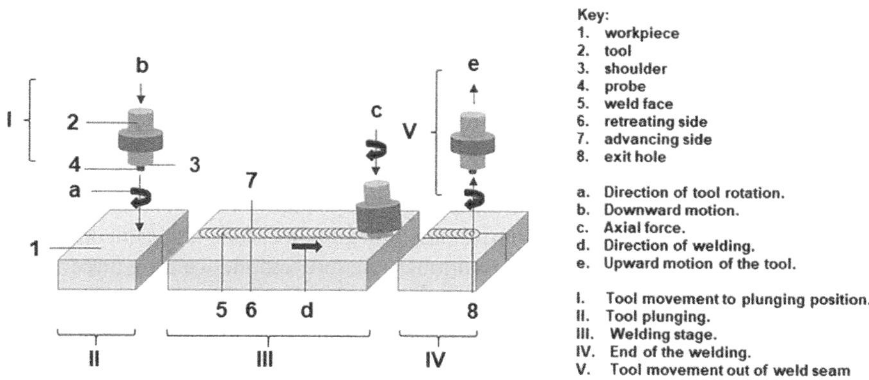

Fig. 1 FSW operation principle and process stages

addition to the tool plunging a dwell time of the tool can be preset to generate a sufficient heat input before the beginning of the welding stage (III). During the welding stage the plasticized workpiece material intermixes by shoulder and probe and creates a firm bond along the welding path. At the end of the welding process (IV) the tool movement stops first while the shoulder and the probe are still in rotation for a few seconds. Then, the tool is moved out of the joining area by an upward motion. The quality of an FSW joint is determined by a sufficient material flow under and around the tool which requires in particular correct process and tool conditions. With respect to the process conditions, suitable welding settings such as axial force, rotational speed, and tool angle must be preset. Due to a sufficient material intermixing the tool conditions are relevant as well and, therefore, most of the time probes and shoulders are equipped with a specific geometric profile. However, probe and shoulder profiles are affected by tool wear and reduce the material flow during the welding procedure. As the consequence of inappropriate process and tool conditions weld seam irregularities can occur due to insufficient material flow (Mishra and Ma 2005). Therefore, industrial challenges such as the detection and prevention of weld seam irregularities are highly relevant, as the welding area is under the tool and not directly visible.

Usually, the weld seam quality is checked after the manufacturing process by means of various methods (post-process). Thus, weld seam irregularities are detected subsequent to the joining process and result in cost-intensive reworkings or component failure. In particular, suitable non-destructive methods for process monitoring are required to detect weld seam irregularities during the welding process (in-process). Previous studies have examined a variety of approaches to implement suitable process monitoring methods, whereby measurable signals from the joining process or the welding setup are used. These include mechanical loads such as forces and torque, power and current signals of the spindle, temperature in the joining area, vibrations, as well as audio signals. In particular the interactions between audio signals and weld seam irregularities were widely investigated so far.

Chen et al. (2003) used structure-borne sound measurements to monitor the FSW process. Breakouts of different sizes were provided in the weld seam, which served as artificially generated irregularities. Due to the use of wavelet transformations of the structure-borne sound signals, the gaps between the shoulder and the workpiece could be detected, which the authors considered to be a basic possibility of process monitoring. Due to this sensitive measurement method, differences in the audio signals between the advancing and retreating side of the tool can also be detected. This can indicate different flow patterns and deformations of the material, due to the differences between the direction of rotation of the tool and the welding direction of the advancing and retreating side of the tool (Chen et al. 2003).

Soundararajan et al. (2006) investigated the approach of an acoustic process monitoring for FSW as well. It could be shown that the audio signals deviate due to the variation of the rotational speed. If the thermomechanical conditions remain almost constant during joining, the FSW process reaches a steady state. If changes occur during this phase, then the quality of the weld is affected. The interactions of the tool with the workpiece can directly relate to important features in the audio signals. The signals were evaluated with the help of Fast Fourier-, Short Time Fourier- and Wavelet transform. Two significant frequency ranges were found for the FSW, which can be influenced by varying the welding parameters. Since many mechanisms are involved in the generation of the structure-borne sound, the extraction of the significant features becomes difficult (Chen et al. 2003). Due to their sensitivity and fast reaction to process irregularities, the mechanical variables are best suited for real-time monitoring according to Mishra et al. (2018). Force signals were analyzed for the first time as early as 2006 by Boldsaikhan et al. (2006) using artificial neural networks (ANN). Their results show that there is a correlation of the welding quality with the feed and shear forces, as well as the feedback moments. In another study, the authors state that the classification of welds with tunnel defects is possible with 95% accuracy (Boldsaikhan et al. 2011). For this purpose, they use the frequency spectra of the feedback forces generated by means of Fourier transformation and also an ANN approach.

Up to now, airborne sound in the audible range has not been used as a process monitoring in FSW. The reasons for this are the background noises occurs from various production processes and sources in the workshop area. Thus, a separation from the relevant acoustic signals is required. However, due to the advancement of computer hardware and machine learning, especially deep learning, the analysis of audio signals is undergoing a rapid development. This is made possible by the efficient training and autonomous feature extraction from audio data (Chan and Chin 2020; Grollmisch and Johnson 2020, Johnson and Grollmisch 2020).

Several publications have already confirmed the promising potential of airborne sound analysis for process monitoring in welding procedures. Suitable machine learning algorithms for the evaluation of airborne sound enable the detection of different weld seam irregularities with detection probabilities up to 90% (Sumesh et al. 2015; Ren et al. 2018; Asif et al. 2020). Furthermore, microphones are comparatively not as expensive and can be installed very close to the joining process. This provides in particular a wide range of applicability for large-scale components due to the absence of local fixed structures which are used for structure-borne sound or in local force measurement devices.

Due to the existing potential this study focuses on the basic application of airborne sound as a promising method to detect weld seam irregularities during the FSW process. Initially, a method was developed to evaluate the correct mounting position of a suitable microphone. Then it is shown that the varying FSW process stages (see Fig. 1) are recognizable in the airborne sound signal. Furthermore, an analysis of airborne sound and parallel of the axial force F_z was carried out in order to evaluate the significance of both signals in each of the process stages I through V. Finally, it was shown that it is indeed possible to use airborne sound measurements to detect weld seam irregularities.

2 Experimental Design

The experimental design considered (a) the directional characteristics of the process ($+x$, $-x$, $+y$, and $-y$) and (b) the position of the microphone. Prior to the measurements the spatial sound radiation characteristics of FSW processes were unknown. Therefore, a survey of various microphone positions was necessary.

2.1 General Conditions

The experimental investigations were conducted on a robotized force controlled FSW setup from Grenzebach Maschinenbau GmbH using a KUKA KR500 heavy duty robot with an axial force of up to 10 kN and a maximum rotational speed of up to 14,000 1/min (Fig. 2).

The welding tool, consisting of a shoulder and a probe, was provided by RRS Schilling GmbH. Figure 3 depicts the welding tool in (a) top and (b) side view position.

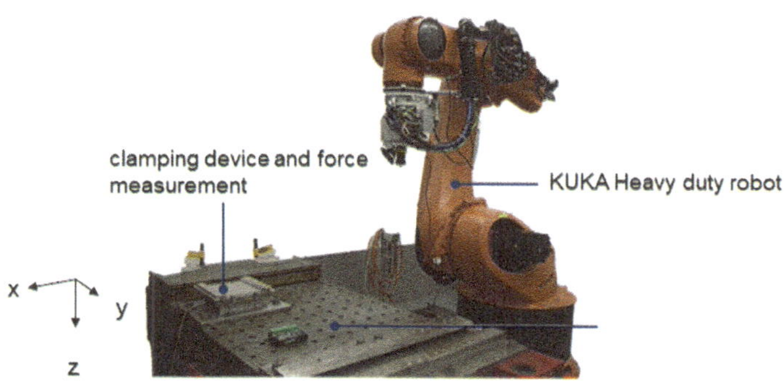

Fig. 2 Robotized FSW setup, clamping device with force measurement unit, and welding table

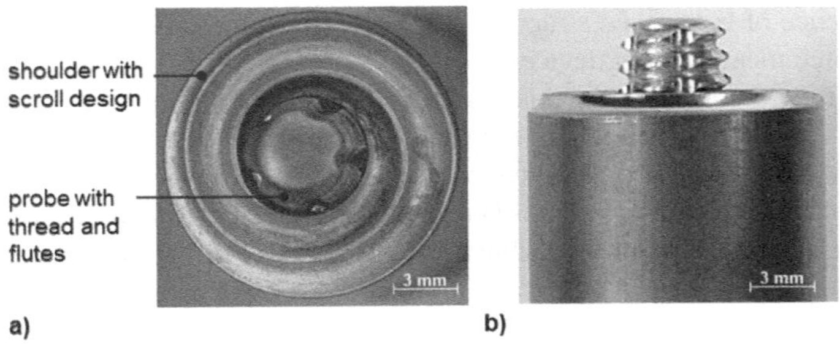

Fig. 3 **a** Top view and **b** side view of the welding tool consisting of shoulder and probe

The shoulder has a diameter of 15 mm and is additionally designed with a scroll profile to improve the material flow whereas the probe was manufactured with a thread and five flutes to ensure a sufficient material intermixing. Furthermore, a probe length of 3.75 mm was preset to prevent a contact with the backing plate. All experiments were conducted at a welding speed of 1000 mm/min, a tilt angle of 2°, a tool rotational speed of 1500 1/min, and an axial force of 8500 N.

The initial experiments (see Sects. 3.1 and 3.2) to investigate the airborne sound emissions during the FSW process were carried out using sheets of the aluminum-magnesium alloy AA 5754 H111 with 5 mm sheet thickness, 250 mm length and 250 mm width. For all weld seams in Sect. 3.1, a length of 50 mm was selected. During the final experiments in Sect. 3.3 a weld seam length of 600 mm and a sample width of 170 mm was used to measure the airborne sound signal during the flash formation in the weld seam (see Fig. 9).

2.2 *Measurement Setup*

For the acoustic measurement, the spindle of the robot is equipped with a condenser microphone sE8 from sE Electronics International, Inc. The microphone has a frequency range from 20 to 20,000 Hz, similar to the human hearing range (sE Electronics International 2021). It has a cardioid characteristic, which means that signals from the front are perceived more strongly than from the rear. In this way it is attempted to only record the main sound of the process and to significantly suppress ambient noise. The measurement setup for the recording of airborne sound during the FSW process is shown in Fig. 4.

To protect the microphone from vibrations and thus avoid distortion of the signals, the microphone is mounted to the robot via elastic microphone shock mounts. The microphone and the mount are attached to the spindle via an adapter plate. The designed adapter plate provides ten different microphone positions (in 36° steps) in relation to the robot coordinate system, so the sound radiation pattern during the process can be investigated in detail. In Fig. 5a the microphone positions are

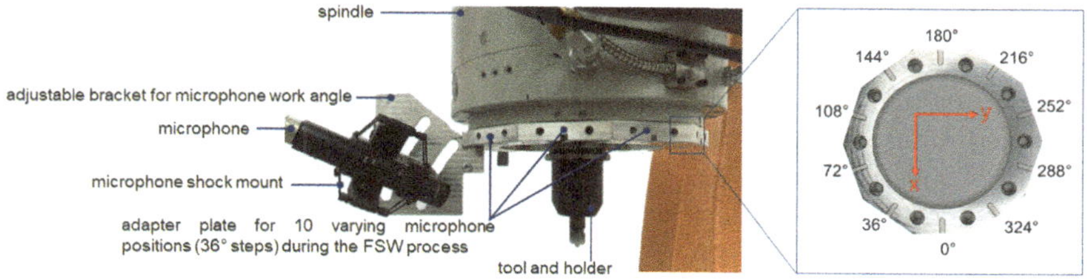

Fig. 4 **a** Measurement setup for the recording of the emitted airborne sound and **b** microphone positions which can adjust by the adapter ring

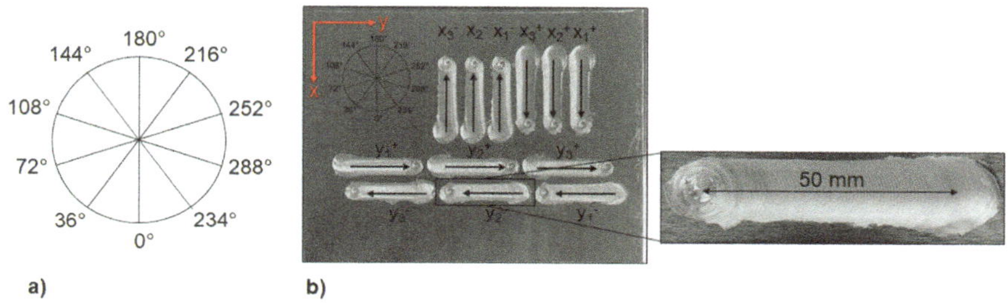

Fig. 5 **a** Varying rotation angles and **b** FSW weld seams joined in $\pm x$ and $\pm y$ direction with an enlarged section of a weld seam

illustrated with varying rotation angles.

During the experiments the spindle moves only in directions of $+x$, $-x$, $+y$, and $-y$ whereby the position of the coordinate system keep constant. Furthermore, an adjustable bracket can be used for work angle settings of the microphone relative to the spindle axis (25–40°). However, the following investigations were performed with a constant angle of 25°. The distance between the microphone and the tool center point remains constant at 150 mm.

The microphone is connected to the required 48 V phantom power supply via an XLR cable whereas the data were recorded with a sample rate of 100 kHz. The process force F_Z was measured with a multi-component dynamometer type 9255C by Kistler with a maximum scan rate of 50 kHz, mounted below the clamping device (see Fig. 2). Both the microphone and the force measurement device are connected with a Dewetron measuring system to ensure the synchronization of the audio and force signals.

2.3 Experiment Execution

Weld geometries are composed of meanders and curved parts that affect the orientation of the welding tool and, hence, the direction of the sound radiation. Consequently, the positioning of the microphone could be crucial for picking up audio signal. To verify this, the directional characteristics of the process are assessed at various points in order to obtain those positions at which the strongest sound signal (main lobe) is emitted. Consequently, several positions should be observed at the same time, if necessary. Spherical directivity at each measured point enables numerous advantages to the practicality of acoustic condition monitoring.

The experimental investigations were carried out at TU Ilmenau. The first analyses focused on the FSW process. Therefore, the sealing air (prevent interior spindle soiling) was switched off because it could override the welding process acoustically. Figure 5b shows the 50 mm long weld seam geometries that were recorded acoustically; three for each direction (+*x*, −*x*, +*y*, and −*y*). This results in a data set of twelve recordings for each microphone position.

For further consideration, the mean of the three individual welds (of each direction and microphone position) is analyzed for its frequency and its amplitude. These are then plotted in a polar plot to assess the frequency response as a function of position and amplitude.

3 Results and Discussion

Before the audio signal could be interpreted, the directionality of the FSW process acoustic emissions was analyzed. This was necessary to determine the direction from which the welding robot emitted the audio signal. The directionality was analyzed by polar plots.

3.1 Acoustic Characterization by Polar Plots

As already shown polar plots are used to evaluate the directionality; they show which frequencies are suppressed or emphasized in a particular direction. To create the polar plots, the axial force signal was searched for zeros. The first change of sign in the second half (back area of the weld) of the recording (equal to first zero point) marks the upward motion of the tool at the end of the FSW process detaching from the workpiece (see Fig. 1 IV). Starting from this point, the welding phase was determined backwards. From the recorded audio data, a time window of stage III (see Fig. 8) was defined for each weld sample and considered in the evaluation. Within the Fourier transform (FFT/DFT) and octave filters the data sample was examined. Subsequently, the arithmetic mean of the three welds per direction and microphone position were determined accordingly. The values were plotted in polar diagrams in steps of 36°.

Figures 6 and 7 show the polar plots created for the four different welding directions. Sound pressure levels at higher frequencies are not shown here as the relevant information is located in the low-frequency range.

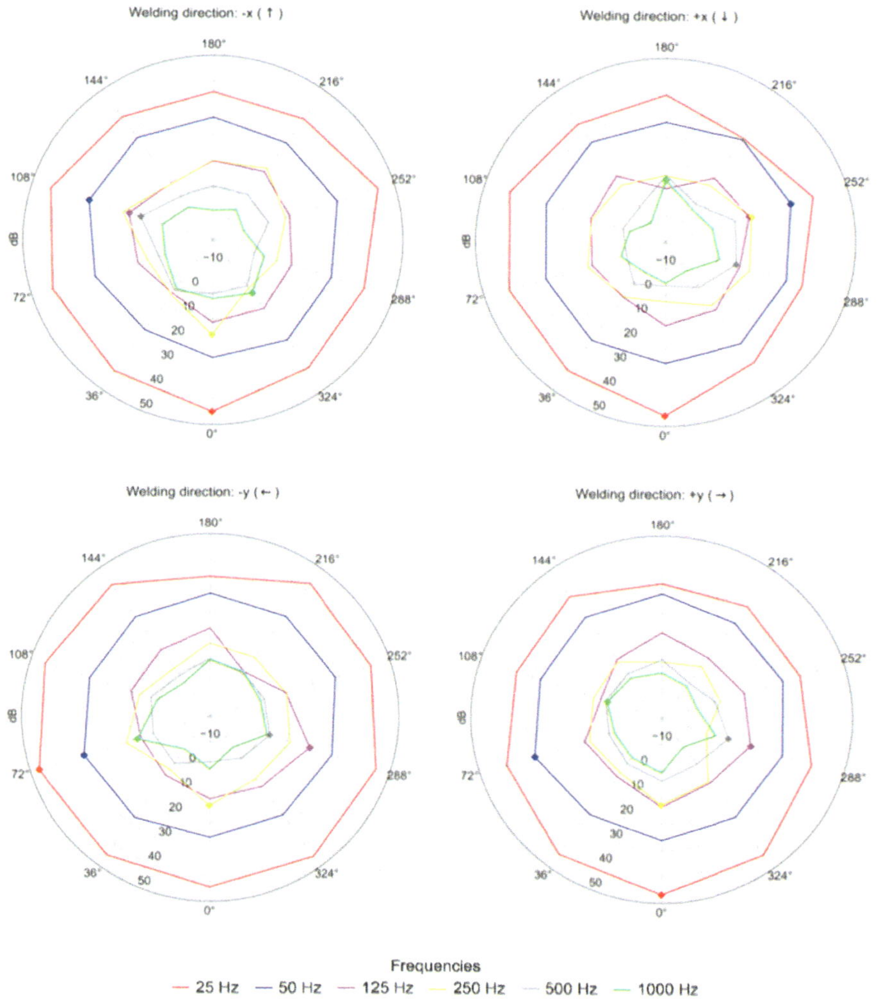

Fig. 6 Polar plots of **a** $+x$ direction, **b** $-x$ direction, **c** $+y$ direction, and **d** $-y$ direction

The polar plots show that relevant frequency levels are equally attenuated in all directions. No significant differences between the process directions $+x$, $-x$, $+y$, and $-y$ were recognized. So, the directionality of the FSW process has no direct preferred direction. Consequently, the position of the microphone can be chosen as needed. This is a particular advantage for the use of acoustic condition monitoring in a real production environment.

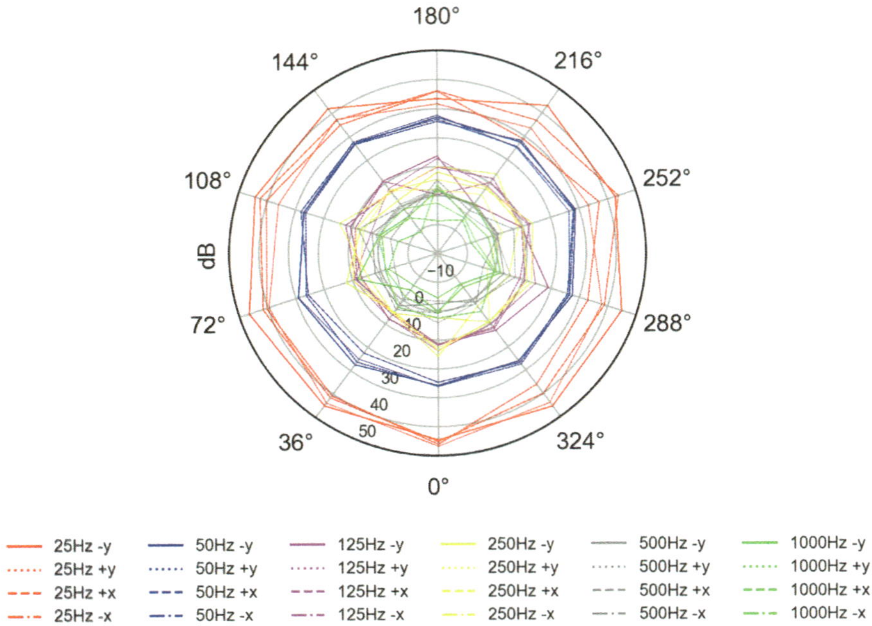

Fig. 7 Comparative polar plot of selected frequencies as a function of all directions

3.2 *Comparison of Process Forces and Audio Signal*

After the investigation of the microphone position, the audio signal was analyzed with respect to its possible significance for acoustic condition monitoring. This was done by means of the acting force in *z*-direction (F_z) and the amplitudes of the audio signal during the joining process. Figure 8 depicts F_z, sound pressure and spectrogram of the audio signal.

Initially, a 50 mm weld seam without irregularities (Fig. 5b) was produced by using an axial force of 8500 N, a rotational speed of 1500 1/min and a welding speed of 1000 mm/min. In stage I (0–2.3 s) the rotating tool moves to the start position of the weld seam. In this stage there is no axial force F_z as the tool and the force measurement setup are not in contact with each other. The microphone signal also shows no significant sound pressure differences. During the second process stage (II) the tool plunges in the workpiece and plasticizes the material due to the frictional heat. In the range between 2.3 s and 12.4 s the axial force F_z increases from 0 up to 8500 N which was the preset value in the force control settings. Thus, the unequal force amplitudes clearly indicate unstable tool/workpiece interaction. After reaching 8500 N the tool plunging is finished and the force control software changes to the next control cycle in stage III. The delay caused by the switch from stage II to stage III is recognizable between 11.2 and 12.4 s and causes a short-term decrease of the axial force from 8500 to 4000 N. Compared to the F_z force signal, the audio signal

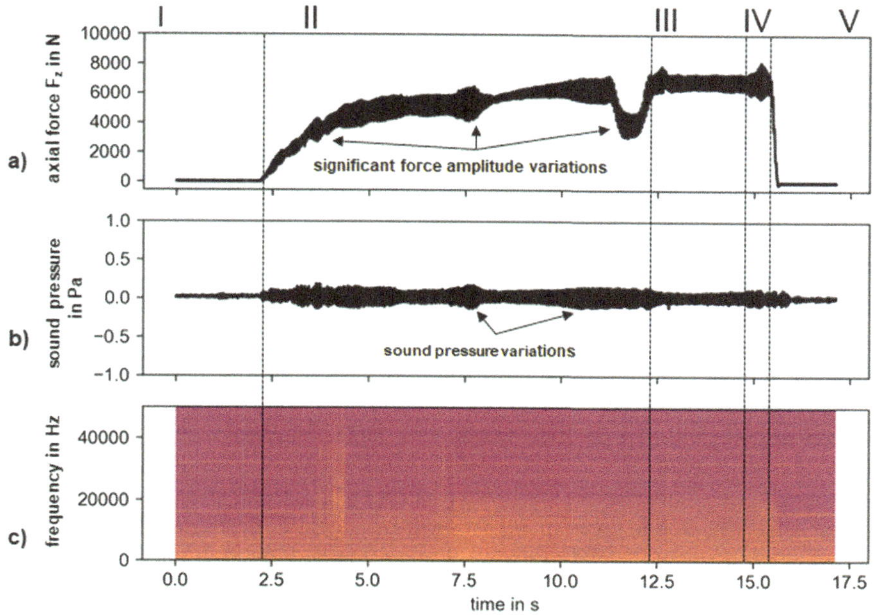

Fig. 8 **a** Axial force F_z, **b** sound pressure, and **c** spectrogram of the audio signal as a function of time and process stages I–V

depicts only a very slight increase of the sound pressure in this range (see Fig. 8b). In the welding stage (III) the force and audio signals are approximately constant, which indicates a stable process performance without weld seam irregularities. Subsequent to the welding stage (14.7–15.3 s) the tool stops but is still in rotation for a few seconds (IV). At the end of the joining process the axial force and the tool rotation decreases to 0 N and 0 1/min and the tool moves to the initial position (V). The audio signal in IV and the beginning of V exhibits a slight increase of the amplitudes as long as the rotating tool and the workpiece material are still in contact. In stage V (16–17.2 s) the audio signal is approximately at 0 Pa.

3.3 Detection of Weld Seam Irregularities Using Audio Signals

As mentioned above, axial force and audio signal were compared to verify the general applicability of the FSW process. However, it still needs to be shown whether it is possible to detect weld seam irregularities by means of an analysis of the emitted airborne sound. Therefore, a 600 mm weld seam was fabricated to investigate the

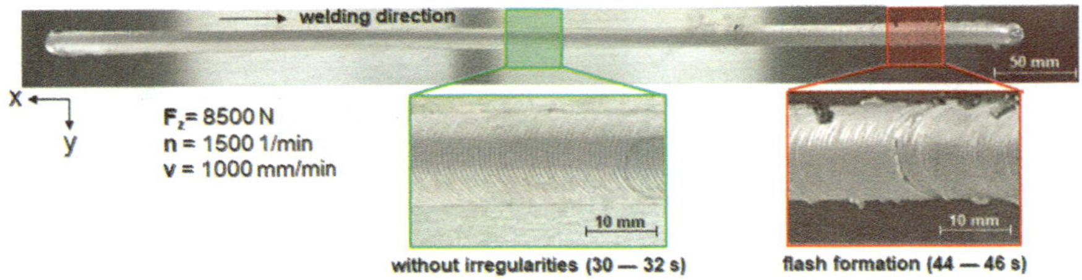

Fig. 9 Weld seam with varying surface quality at the center and at the end

occurrence of surface irregularities such as flash formation. Figure 9 shows the weld seam surface with varying surface conditions.

The weld seam shows varying surface conditions although constant process settings (F_z = 8500 N, n = 1500 1/min, v = 1000 mm/min) were preset. A major part of the weld seam shows a surface with the characteristic extrusion profile of the material displacement during the FSW process, which is for the most part free of irregularities (marked in green in Fig. 9). However, areas of flash formation (44–46 s) were also observed at the end of the weld seam (marked in red in Fig. 9). The audio signal of irregularity-free weld and flash formation is illustrated in Fig. 10a.

Due to the fact that weld seam irregularities mostly occur during the welding stage (III), the following explanations refer to the period from 14 to 48 s. Therefore, two selected areas are considered where a defect-free surface (30–32 s) and a surface with flash formation (44–46 s) can be observed. In the area without weld seam irregularities (see Fig. 10b), an almost regular audio signal was noticed (30–32 s). The amplitudes of the sound pressure depict no significant variations. However, the area with flash formation (see Fig. 10c) exhibits irregular audio signal with sound pressure variations. This can be explained by deviating surface properties (contamination, oxide layer thickness) or insufficient material homogeneity. So the undesired interactions between tool and material, resulting in flash formation, seam detectable.

Figure 11 shows the spectrogram of an FSW weld seam with the corresponding recorded acoustic time signal. It can be seen that irregularities in the weld seam show changes in the audio signal. Irregularities in the FSW process lead to frequency fluctuations or shifts.

This further confirms that flash formations can potentially be detected by means of acoustic measurements and analyses. However further experiments leading to a higher amount of experimental data which are needed to verify this hypothesis.

4 Summary

In this study airborne audio signals were considered for the detection of irregularities during the FSW process. It could be shown that acoustic condition monitoring

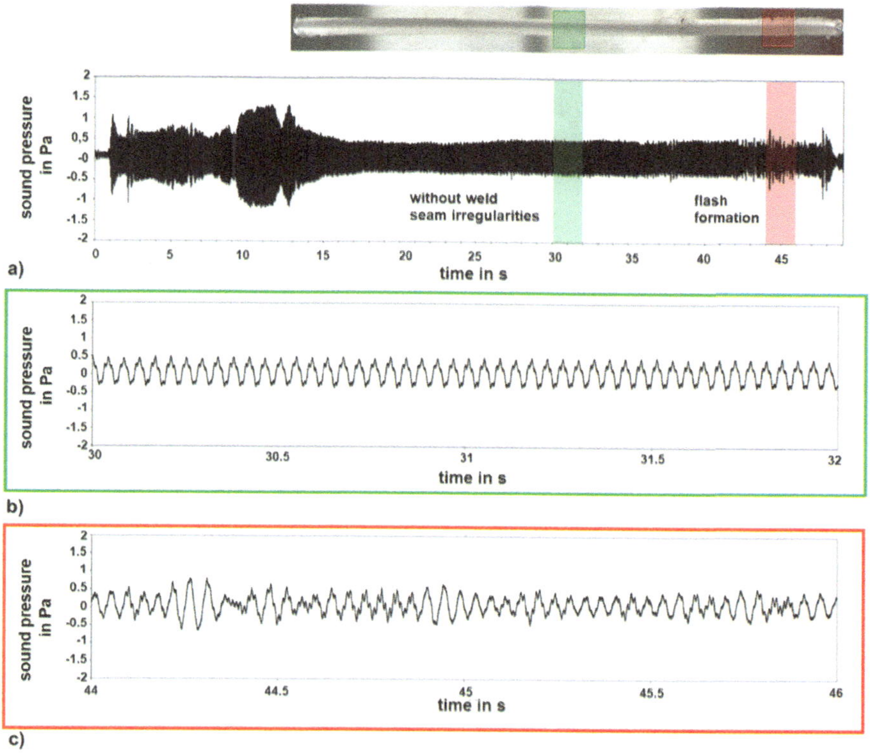

Fig. 10 **a** FSW weld seam and sound pressure. **b** Without weld seam irregularities and **c** during flash formation

has a high potential and could be a good addition or alternative to other measurement methods. The acoustic condition monitoring with airborne sound offers some advantages, in addition to non-destructive and non-contact measurement.

In Sect. 3.1 of this study the measurement position of the microphone was investigated with varying welding directions and at various angles related to the welding direction. This was carried out to determine a suitable position for the microphone mounting during the FSW process. Then in Sect. 3.2 a comparison between the audio signal and the acting process force F_z were carried out to detect weld seam irregularities in general. Finally, in Sect. 3.3 audio emissions were used for the detection of flash formation during the joining process. All experiments were carried out with 5 mm aluminum sheets (AA 5754-H111) and a weld seam length of up to 600 mm. In summary, the following conclusions can be made:

- The positioning of the microphones does not require a preferred direction in front of, behind, or to the side of the welding process. Furthermore, it was shown that the welding direction does not affect the microphone position. This was verified by means of the cardioid characteristic of the microphone. Thus, free mounting

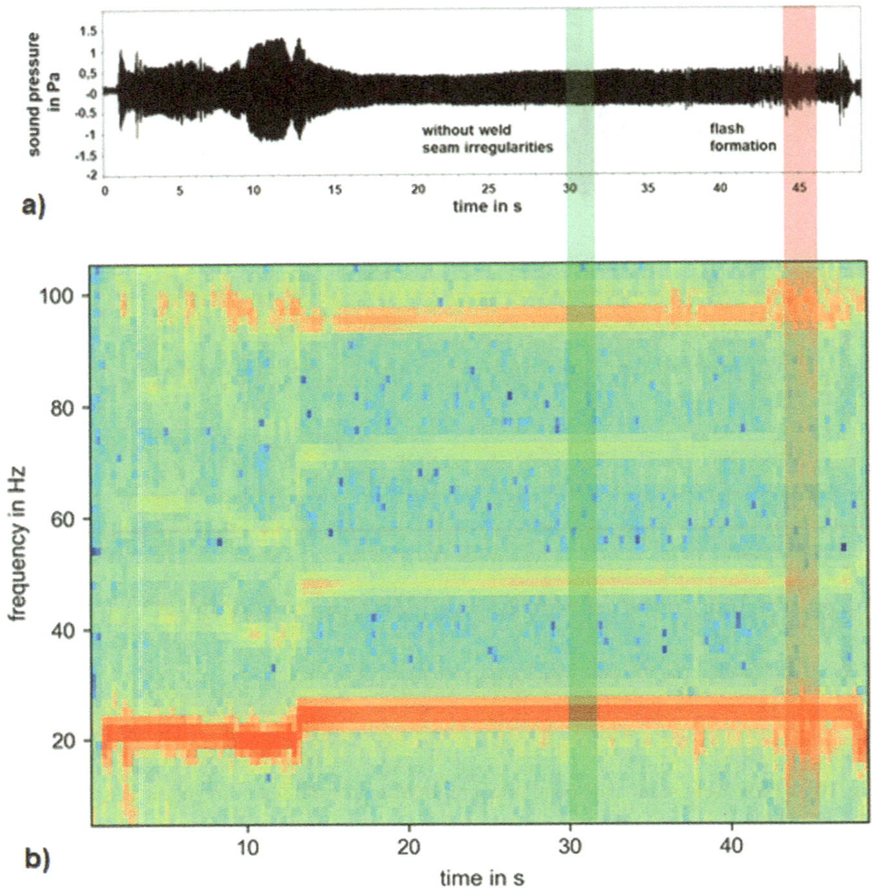

Fig. 11 **a** Time signal of welding seam and **b** corresponding spectrogram

positions of the airborne sound microphone during the FSW process could be proven.

- The general significance of airborne sound signals was observed by an analysis between the axial force F_z and the audio signal. It was shown that relevant FSW process stages are detectable in the airborne sound signal. Furthermore, significant variations in the force signal can also be detected in the audio signal.
- The FSW setup-related effects are also detectable in the airborne signal. The transition zone between tool plunging and welding stage is detectable as well. Thus, airborne sound can be used for an optimized process handling without non-desirable delays between the process stages.
- Finally, the general applicability of airborne sound signals for detecting weld seam irregularities was investigated. It was observed that the sound pressure in the audio signal is irregular during the flash formation. Whereas an audio signal of

a weld without flash formation depicts an approximately regular sound pressure without variations. Thus, the observation of the amplitudes and the periodicity of the airborne sound signal represents a promising approach for the detection of weld seam irregularities.

5 Future Work

The application of airborne sound measurements for the detection of weld seam irregularities in FSW requires further detailed investigation of the surface quality and the airborne sound. This means in particular the interactions between the measured frequencies, their intensity as well as the resulting weld seam conditions (defect-free surfaces or surfaces with flash formations). Furthermore, the applicability for other irregularities such as wormhole formation should be investigated in the future.

However, the most relevant challenge, particularly for industrial applications, is the separation of background noise from the production environment, from the joining process, and from the FSW setup. Machine learning methods and algorithms are becoming increasingly suitable for this purpose.

Acknowledgements This research and development project is funded by the German Federal Ministry of Education and Research (BMBF) within the "Anwendung von Methoden der Künstlichen Intelligenz in der Praxis" Program (funding number 01IS20001F) and implemented by the Project Management Agency (PT-DLR) at the DLR (German Aerospace Center). The author is responsible for the content of this publication.

SPONSORED BY THE

Federal Ministry of Education and Research

References

Asif, K., Zhang, L., Derrible, S., Indacochea, E., Ozevin, D., Ziebart, B.: Machine learning model to predict welding quality using air-coupled acoustic emission and weld inputs. J. Intell. Manuf. 604 (2020). https://doi.org/10.1007/s10845-020-0166

Boldsaikhan, E., Corwin, E., Logar, A., Arbegast, W.: Neural network evaluation of weld quality using FSW feedback data. In: Proceedings of 6th International Friction Stir Welding Symposium, Saint-Sauveur, Nr Montreal, Canada, 10–12 October 2006

Boldsaikhan, E., Corwin, E.M., Logar, A.M., Arbegast, W.J.: The use of neural network and discrete Fourier transform for real-time evaluation of friction stir welding. Appl. Soft Comput. **11**(8), 4839–4846 (2011). https://doi.org/10.1016/j.asoc.2011.06.017

Chan, T.K., Chin, C.S.: A comprehensive review of polyphonic sound event detection. IEEE Access **8**, 103339–103373 (2020). https://doi.org/10.1109/ACCESS.2020.2999388

Chen, C., Kovacevic, R., Jandgric, D.: Wavelet transform analysis of acoustic emission in monitoring friction stir welding of 6061 aluminum. Int. J. Mach. Tools Manuf. **43**, 1383–1390 (2003). https://doi.org/10.1016/S0890-6955(03)00130-5

Grollmisch, S., Johnson, D.S.: Visualizing neural network decisions for industrial sound analysis. Meas. Sci. Technol. (SMSI), Nuremberg, Germany, 267–268 (2020). https://doi.org/10.5162/SMSI2020/D2.2

Johnson, D.S., Grollmisch, S.: Techniques improving the robustness of deep learning models for industrial sound analysis. Eur. Signal Process. Conf. (EUSIPCO), Amsterdam, Netherlands, 81–85 (2020). https://doi.org/10.23919/Eusipco47968.2020.9287327

Mishra, R.S., Ma, Z.Y.: Friction stir welding and processing. Mater. Sci. Eng. R Rep. **50**(1–2), 1–78 (2005). https://doi.org/10.1016/j.mser.2005.07.001

Mishra, D., Roy, R.B., Dutta, S., Pal, S.K., Chakravarty, D.: A review on sensor-based monitoring and control of friction stir welding process and a roadmap to Industry 4.0. J. Manuf. Process. **36**, 373–397 (2018). https://doi.org/10.1016/j.jmapro.2018.10.016

Ren, W., Wen, G., Liu, S., Yang, Z., Xu, B., Zhang, Z.: Seam penetration recognition for GTAW using convolutional neural network based on time-frequency image of arc sound. IEEE 23rd IEEE Int. Conf. Emerg. Technol. Fact. Autom. ETFA, 853–860 (2018). https://doi.org/10.1109/ETFA.2018.8502478

sE electronics international: user manual sE8. Novato, 2021

Soundararajan, V., Atharifar, H., Kovacevic, R.: Monitoring and processing the acoustic emission signals from the friction-stir-welding process. Proc. Inst. Mech. Eng. B. J. Eng. Manuf. **220**, 1673–1685 (2006). https://doi.org/10.1243/09544054JEM586

Sumesh, A., Rameshkumar, K., Mohandas, K., Shyam Babu, R.: Use of machine learning algorithms for weld quality monitoring using acoustic signature. Procedia Comput. Sci. **50**, 316–322 (2015). https://doi.org/10.1016/j.procs.2015.04.042

Influence of Tool Wear on Weld Quality in Refill Friction Stir Spot Welding of Aluminium

D. Lauterbach, D. Keil, A. Harms, M. Schulze, and K. Dilger

Abstract As reported in numerous studies, tool wear is a significant challenge in bringing Refill Friction Stir Spot Welding (RFSSW) into series production. The known studies investigate the influence of the welded material on the tool wear behaviour. Approaches have been made, applying wear resistant coatings in order to extend tool life, but have not taken into account the weld quality. This study aims on defining quality criteria of a typical spot weld produced by RFSSW and to analyse how they are affected by the tool wear. Therefore, two tool sets with different wear resistant coatings are investigated, which show a different wear behaviour. In order to characterize the weld quality, microstructural as well as mechanical properties of the welds are considered. Cross-sections and microstructural analysis indicate the impact of differences in temperature evolution during welding on the weld's microstructure. A comparison is set up between both coatings at different stages of tool life. Additionally, parameter studies determining cross and shear tensile strength at each stage show the variation of optimized welding parameters according to the occurring wear mechanisms on the tool. Thus, the different frictional conditions of the coatings over the tool life as well as the geometrical deviations of the tool sets are considered, in order to determine their influence on the weld quality. The results of this research help to understand the influence of frictional behavior and

D. Lauterbach (✉) · A. Harms · M. Schulze
Volkswagen AG, Group Innovation, Berliner Ring 2, 38440 Wolfsburg, Germany
e-mail: dennis.lauterbach@volkswagen.de

A. Harms
e-mail: alexander.harms@volkswagen.de

M. Schulze
e-mail: maik.schulze2@volkswagen.de

D. Keil
Volkswagen AG, Technical Development, Berliner Ring 2, 38440 Wolfsburg, Germany
e-mail: daniel.keil@volkswagen.de

K. Dilger
Institute of Joining and Welding, TU Braunschweig, Langer Kamp 8, 38106 Braunschweig, Germany
e-mail: k.dilger@tu-braunschweig.de

L. F. M. da Silva et al. (eds.), *2nd International Conference on Advanced Joining Processes (AJP 2021)*, Proceedings in Engineering Mechanics,
https://doi.org/10.1007/978-3-030-95463-5_4

geometrical deviations of the tool on the weld quality and the mechanical properties of the resulting weld.

Keywords Refill friction stir spot welding · Tool wear · Wear-resistant coating · Weld quality

1 Introduction

The various studies on refill friction stir spot welding (RFSSW) demonstrate a wide range of applications reaching from the fields of aviation and aerospace engineering to the transport sector and automotive production. (Fukada et al. 2013; Goushegir 2015; Xu et al. 2018). In particular, the joint formation and microstructural as well as mechanical properties have been addressed by several recent studies (Castro et al. 2018; Amancio-Filho et al. 2011; Cao et al. 2016; Tier et al. 2017; Camilo et al. 2014; Kubit et al. 2018; Hovanski et al. 2021). However, the minority of studies published to date have addressed the tool wear, due to which the technology has not reached a widespread use in a series production e.g. in the automotive industry yet. The high tool costs combined with the limited tool life make its use unattractive at present. Therefore, a few studies focus on investigations regarding tool life and tool wear behavior during RFSSW. In SLV München (2010), 2.0 mm EN AW 5454 sheets were joined and two independent tool sets remained intact after a total of 9000 spot welds each. During the tests, a sodium hydroxide solution was used to clean the tools every 600 welds, which kept the tools smooth-running and allowed the wear conditions to be monitored. For other investigations in Montag et al. (2014), 2.0 mm EN AW 6082 sheets are welded with initial turns of 70 welds, later 210 welds, after which the tool was cleaned in sodium hydroxide as well. During the breaks, the tool condition was monitored and after approximately 3100 welds, the tests were terminated due to a worn out tool. With an increasing number of welds, an increasing tool wear led to an increase regarding the spindle current. Therefore, a higher torque during the process was indicated. During tool monitoring, the fastest wear occurred on the sleeve, showing different wear rates in different areas. Especially around 2 mm above the front face of the sleeve, the diameter decreased with the highest rate. Therefore, more and more material penetrated the tool, increasing the tool temperature and thus enhancing the aluminium absorption, which was assumed as a self-reinforcement of the tool wear. The most severe wear effects were observed in Nasiri et al. (2018), as pin and sleeve glowed after only 5 welds on Lithium-alloyed EN AW 2099 sheets. Then, after a short cooling time, the tool was jammed so tightly, that it could no longer be rotated. Microstructural changes as well as severe erosion at the sleeve surface and chipping at the front face of pin and sleeve were detected. Own investigations published in Lauterbach et al. (2021) show a significant improvement of the tool wear behaviour by using wear resistant PVD coatings. During the welding tests, the tools were operated at temperatures of up to 550 °C, which led to severe changes regarding the mechanical and tribological properties of the coatings. A high hardness in combination with a high adhesive strength on CrVN coated tools led to a significant reduction in tool wear. As both, hardness and adhesive strength, also remained on a high level after a temperature treatment of 300 min at 600 °C, the tool set reaches 6400 spot welds and remains intact, producing sound welds. CrVN is a state of the art coating for casting tools, especially in the field of aluminium and magnesium die casting. Its vanadium content forms an oxide on the surface above approx. 550 °C, which reduces its coefficient of friction (CoF) (Paschke et al. 2010). Cr-DLC is used as the second coating, which is known from several applications for cutting tools. Both coatings are applied on a Cr adhesive promoting layer. The focus of this study is to investigate the influence of the wear-related changes of the tool regarding its geometry as well as the tribological properties on the welding quality. For this study, the focus is to be set on the welding result. Different studies investigate the principal properties of a joint produced by RFSSW (Castro et al. 2018; Ferreira et al. 2020; Li et al. 2016, 2019; Chai and Wang 2019) and how they are affected by different welding parameters. This study aims on the characterization of the welds using similar methods and the development of quality criteria depending on the tool properties.

2 Methods and Materials

At first, the coatings are characterized and compared according to the methodology in (Lauterbach et al. 2021). Therefore, the geometrical, mechanical and tribological properties are measured using standard methods. Table 1 gives an overview of the properties and the used characterization methods. The mechanical properties hardness and adhesion strength along with the coefficient of friction are determined in the initial state of the coatings, right after the coating process, as well as after a temperature treatment. The coated specimen are temperature treated at 600 °C for 300 min in order to simulate the temperature exposure during the operation of the tool. The changes during the temperature exposure showed a significant influence on the wear behaviour of each coating. The two different coatings are then each applied on a tool set for refill friction stir spot welding.

From each tool set, pin and sleeve were PVD coated on the outside, as the inside of the sleeve can not be reached by the sputtering process. The tools are used for

Table 1 Coating properties and methods of determination

Properties	Method	Standard
Coating thickness	Calotte grinding	DIN EN 26423
Surface roughness	Confocal microscopy	DIN 4760
Coating hardness	Vickers hardness test	DIN EN ISO 14577-1
Adhesion strength	Scratch-test	DIN EN ISO 20502
Coefficient of friction (Cof)	Scratch-test	DIN EN ISO 20502

Table 2 Welding parameters for the tool life tests (PR = Permanent rotation)

Step	Time	Rotational speed (min^{-1})	Pos. sleeve (mm)	Pos. pin (mm)
0 (PR)	0.0	200	−0.300	−0.500
1	0.3	2200	0.000	0.000
2	0.2	2200	1.200	−1.173
3	0.6	2200	−0.307	0.300
4	0.2	2200	0.000	0.000

durability tests on aluminium sheets in order to analyse their specific behaviour during the welding process. The welding tests are carried out on a "RPS 100" from Harms & Wende GmbH (Hamburg, Germany), which is modified for welding experiments with one-sided accessibility. The welding parameters were chosen from previous studies and represent a welding programme designed for a short welding time. The short welding time enhances the wear effects and increases the wear rates. The welding programme is given in Table 2.

According to the table, the total welding time is 1.3 s with a plunging depth of 1.2 mm. As given in the table, after plunging the sleeve, the pin penetrates the material as well. This step is supposed to improve the connection between the welding zone and the surrounding top sheet, as shown in previous parameter studies. Figure 1 shows the principle of the refill friction stir spot welding process as used in these studies with the four process steps placing the tool on the surface (1), plunging of the sleeve and retraction of the pin (2), retracting the sleeve and plunging the pin (3) and finally setting both to their initial position (4), leaving an even and smooth spot weld. The varied parameters in this study are the rotational speed of sleeve and pin n, plunge depth PD and plunging speed of the sleeve in the second step, v_{plunge}, as well as the correction speed v_{corr} and correction depth of the pin in the third step CD.

The high plunging speed of the sleeve in combination with the short total welding time resulted in critical welding results, as the flank connection of the spot welds was identified as a decisive quality criterion. For the tool life tests, two different material

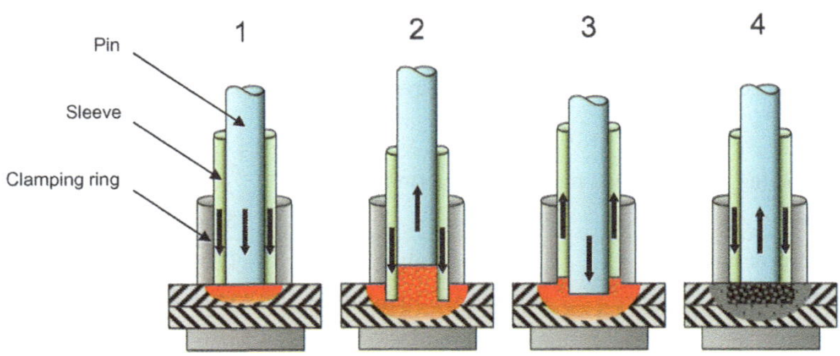

Fig. 1 Schematic principle of the refill friction stir spot welding process

Table 3 Material thickness combinations for the tool life tests

Name	Upper sheet	Thickness (mm)	Lower sheet	Thickness (mm)
MTC 1	EN AW 5083-H111	1.0	EN AW 7020-T6	2.0
MTC 2	EN AW 5083-H111	1.0	EN AW 6082-T6	1.5

thickness combinations (MTC), given in Table 2, were used. Starting with MTC 1, 20 subsequent spot welds were executed, followed by 20 spot welds with MTC 2. The minimum distance between two welds was 75 mm in order to homogenize the heat development of the aluminium sheets (Table 3).

In this configuration, 1600 spots were welded subsequently, before the tool was demounted and analysed. During the welding experiments, the time between two welds reached around 6 s, as one spot weld was started immediately after positioning the sheets. Each 1600 spot welds, the sleeve was investigated using microscopy to determine the aluminium penetration on the sleeve and computer tomography (CT) for the non-destructive analysis of the cross-section. The tool wear was measured on different spots along the thread on the sleeve, where the most severe wear occurs, as also reported by Montag et al. (2014). Additionally, the energy consumption needed in order to plunge the sleeve as well as the tool temperature, which is measured on the outer surface of the clamping ring, are taken into account as process parameters.

In a further step, detailed parameter studies are carried out at intervals of 1600 spot welds after removal of the tool and CT analysis, which are used to evaluate the spot welds. For this purpose, design of experiments is used to create a methodology to determine the influences of the different factors. A partial factorial test plan according to Taguchi in the format L25 was chosen, in which the five influencing variables rotational speed, plunge depth, plunging speed, correction speed and correction depth were varied on five levels each. Correction speed and correction depth refer to the speed at which the pin is plunged into the upper sheet in the third step and how far the displacement is. Table 4 shows the parameter limits of the studies. In this way, cross-tensile as well as shear tensile specimens were prepared and tested with the universal testing machine Zwick 100 from Zwick/Roell (Ulm, Germany). Three specimens are manufactured for each test type and tested at a traverse speed of 10 mm/s and the maximum force is evaluated. The parameter studies are carried out with a material thickness combination consisting of an EN AW 6082 cover plate in 1.5 mm thickness and an EN AW 7020 base plate in a thickness of 2.0 mm. In each

Table 4 Overview of the parameter settings for the welding tests

Factor	Minimum	Maximum	Step size
Rotational speed (min^{-1})	1400	3000	400
Plunge depth (mm)	1.4	2.0	0.15
Plunging speed ($mm\ s^{-1}$)	1.0	5.0	1
Correction speed ($mm\ s^{-1}$)	1.0	5.0	1
Correction depth (mm)	0.0	0.4	0.1

case, one additional spot weld is prepared for metallographic analysis. Therefore the welding spots were cut, grinded and polished down to 1 μm and electrochemically etched using tetrafluoroboric acid in order to analyse the microstructure.

3 Results and Discussion

First, the results of the characterisations of the two different coatings are considered. In particular, the hardness, adhesive strength and coefficient of friction (CoF) are relevant for the following considerations. The three measurements are taken in the initial state after coating, as well as after a temperature treatment at 600 °C over a period of 300 min. Table 5 shows the results of the mechanical and tribological tests.

Each value is an average of at least five individual measurements. It can be seen that changes occur due to the effect of temperature. At the beginning, the coating hardnesses are 1523 HV for CrVN and 1237 HV for Cr-DLC. After temperature treatment, there is no longer any significant difference between the two values: 30.3% is lost in the case of CrVN, and only 12.5% in the case of Cr-DLC. In both cases, the adhesion strength of the CrVN is significantly higher than that of the Cr-DLC, while the decrease due to the elevated temperature is particularly high for the Cr-DLC at 61.2%. In contrast to the mechanical properties, the coefficient of friction of the two coatings behaves in opposite directions. The vanadium in the CrVN forms an oxide on the surface above a temperature of approx. 550 °C, which reduces its coefficient of friction. Thus, a drop from initially 0.43–0.39 is observed in the measurements. The Cr-DLC, on the other hand, increases its coefficient of friction from 0.38 to 0.41. The changes in the coating properties are decisive for the investigations, as the tool conditions change in the course of the tool life tests.

Next, the tool life tests are considered. Both tools are welded to 6400 points according to the described procedure. Meanwhile, the process data are recorded, of which the tool temperature on the outside of the clamping ring and the energy required to plunge the sleeve are considered. Both measured variables are shown in the diagrams in Fig. 2.

For the CrVN layer, after the start of the tests, there is a rapid temperature increase from about 250 °C at the beginning to 350 °C after the first 1600 points. Thereafter,

Table 5 Results of the mechanical and tribological characterization of the coatings

Value	CrVN	Cr-DLC
Hardness (HV)	1523	1237
Hardness after T (HV)	1062	1082
Adhesion strength (N)	163	139
Adhesion strength after T (N)	98	54
Coefficient of friction (/)	0.43	0.38
Coefficient of friction after T (/)	0.39	0.41

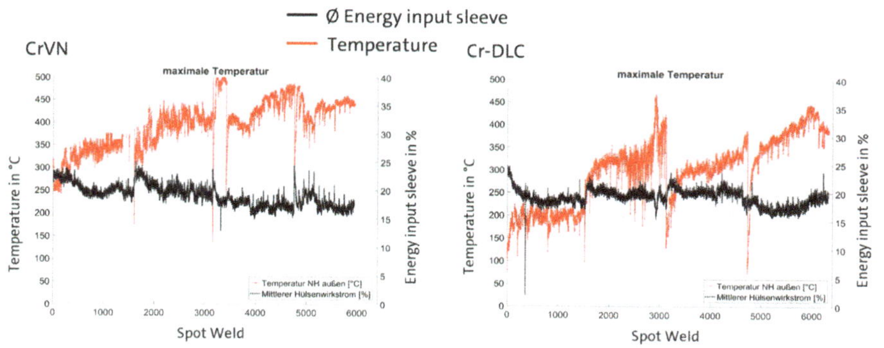

Fig. 2 Process parameters during the tool life tests of CrVN (left) and Cr-DLC coating (right)

the temperature continues to rise, to approx. 400 °C after 3200 points. In the further course, the temperature rises to a maximum of 500 °C, once briefly at 3400 P and once steadily at approx. 5000 P. After 6027 continuous spot welds, the sleeve broke during the process, leading to the termination of the tests with the CrVN tool. When looking at the energy required to plunge the sleeve, a decrease can be seen. As the tool temperature increases over the 6027 points, the energy consumption decreases from an initial 24% to approximately 18%. The increasing tool temperature causes the material to become steadily more plasticized. The resistance of the aluminum against the penetration of the tool decreases, so that it is easier to plunge the sleeve. In comparison, the Cr-DLC tool remains at a very low temperature level at the beginning. It takes the tool around 1800 points to exceed 300 °C. At approx. 2800 welds, the global maximum of 460 °C is reached. Shortly thereafter, at approx. 3000 spot welds, the temperature drops again. Overall, the temperature of the Cr-DLC tool is significantly higher in the second half of the tests. The energy required to move the sleeve is globally decreasing, similar to the CrVN tool. From approx. 25% at the beginning, it decreases to approx. 20% after 6400 spot welds.

From the process data, it can therefore be seen that the CrVN tool is exposed to a significantly higher temperature load throughout the tool life tests. A temperature of over 350 °C is reached shortly after the start of the tests. The Cr-DLC tool, on the other hand, runs most of the time at a rather low temperature and only rarely and for a short time exceeds temperatures above 350 °C; only towards the end of the test series at 5000 points, this mark is exceeded over a longer period. Thus, the coating properties of the two coating systems described at the beginning also change. The CrVN starts to change its mechanical and tribological properties at an early stage. Its hardness and adhesive strength decrease significantly. This is confirmed by the CT images, which are shown in Fig. 3. In contrast, the Cr-DLC coated tool shows stable behavior. The initially low and overall stable temperatures lead to no changes in the cross-section of the tool in the first 4800 weld spots. Only in the CT scan after 4800 welding points, the first slight constrictions become visible on the lateral surface of the sleeve. The end face, on the other hand, changes only negligibly.

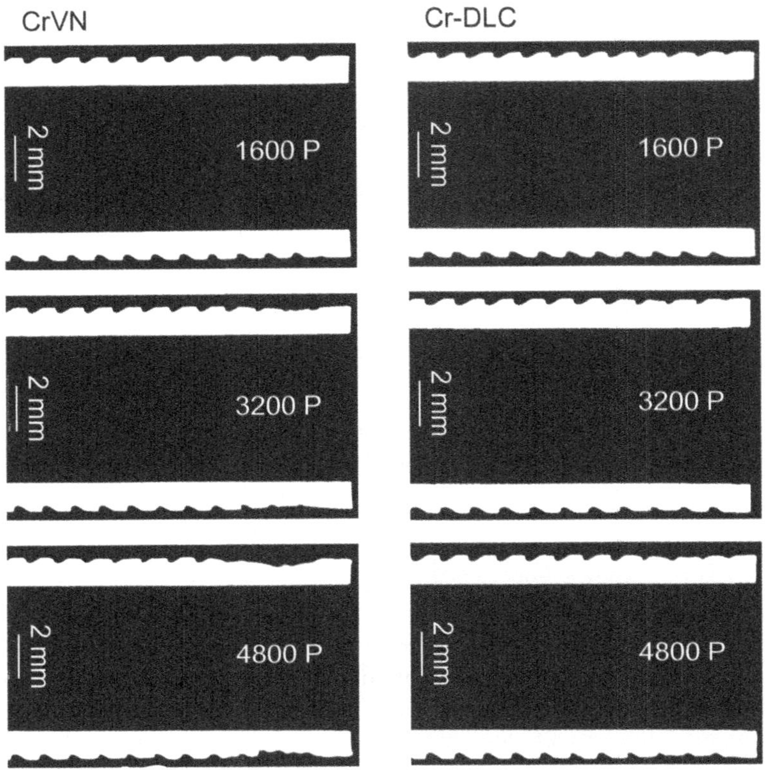

Fig. 3 CT images of CrVN (left) and Cr-DLC tool (right) after 1600, 3200 and 4800 welds

To quantify the wear, the CT scan is used to measure the diameter on the outer surface of the thread on the first nine stages. The result of the two tools is shown in Fig. 3. In each case, the end of the sleeve that is engaged in the workpiece is shown on the right side. In the upper section of the two sleeves, which is not plunged into the material, the sleeve measures 8.75 mm in diameter. In the lower five threads, on the other hand, severe wear is already evident after 3200 spot welds, particularly in the CrVN tool. After 4800 spot welds, the minimum diameter in the first thread is still 8.25 mm. It is noticeable that the highest diameter on the tool is measured at the tip. This is due to aluminum buildup, which can accumulate due to an increased gap between the sleeve and the clamping ring. In the case of the Cr-DLC layer, no geometric changes with respect to the outer diameter of the sleeve are discernible over the entire 4800 spot welds. Here, too, however, buildup can be seen at the tip of the sleeve, where aluminum adheres as a result of the process.

With each tool removal, a parameter study is welded in addition to the CT analysis to investigate the weld quality as well as the welding behavior of the tools. The results of the mechanical investigations show that the parameter sets for sound welds in terms

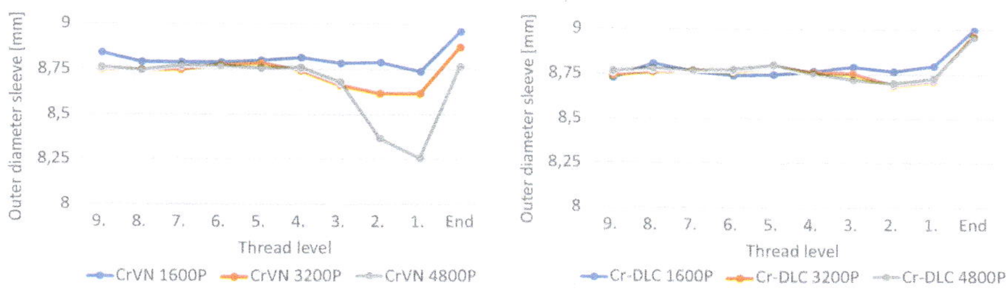

Fig. 4 Outer diameter of each sleeve after 1600, 3200 and 4800 spot welds

of cross and shear tensile strength vary due to changes of the tools. The Cr-DLC tool shows only very slight changes overall, and the welding results are therefore very constant. The results reached up to 9.22 kN regarding the ultimate lap shear strength and 4.32 kN in cross tensile strength. With the CrVN layer, however, significant changes can be recognized. Figure 4 shows the results of the first parameter study in the initial state of the tool and after 1600, 3200 and 4800 welding points as a contour plot. The rotational speed is given on the x-axis and the plunge speed on the y-axis. The cross and shear tensile strength are color-coded. A dark blue represents a low force, while dark green represents the highest strengths. In the first picture, it can be seen that the welding windows for good cross and shear tensile strengths approximately coincide (Fig. 5).

In shear tensile tests, strengths between 3.3 kN at high rotational speeds and high plunge speeds are achieved and up to 8.2 kN at low rotational speed and high plunge speed. The cross tensile strengths are analogously between 0.7 and 4.2 kN in the same parameter windows. In the range of medium plunge speeds, areas with low strengths exist over the complete range of plunging speeds of the sleeve. For the further analyses of joint formation, experiment 10 is selected as a representative welding parameter, which is welded at a rotational speed of 1800 min^{-1} with a plunge depth of 2.0 mm, a plunge speed of 2 mm s^{-1} and no correction depth (0.00 mm). In the initial condition, this exhibits low values in cross and shear tensile tests. After 3200 spot welds, the maximum and minimum values barely change, but the welding ranges where they are reached change. The shear tensile strength has become much more stable, there is no longer a range with forces below 5 kN. In terms of cross tensile strength, however, there are still larger areas with weak welds. Thus, test 10 still achieves a low head tensile strength, but the shear tensile strength increases significantly.

Finally, after 4800 points, a split of the welding window is shown with regard to the cross tensile force. Above a speed of 2200 min^{-1} no higher forces than 2.0 kN are reached. Below that, at least 2.5 kN are achieved with each plunge speed. After a very stable and homogeneous picture at 3200 spot welds, the shear tensile force exhibits a steep sink in the range of 2600 min^{-1} at a plunge speed of 2.0 mm s^{-1}, where only very low forces below 1 kN are reached. Experiment 10 leads to good results in both loading modes. Figure 6 shows the cross-sections of experiment 10 in the initial state of the CrVN tool, after 3200 as well as after 4800 welds. Defects

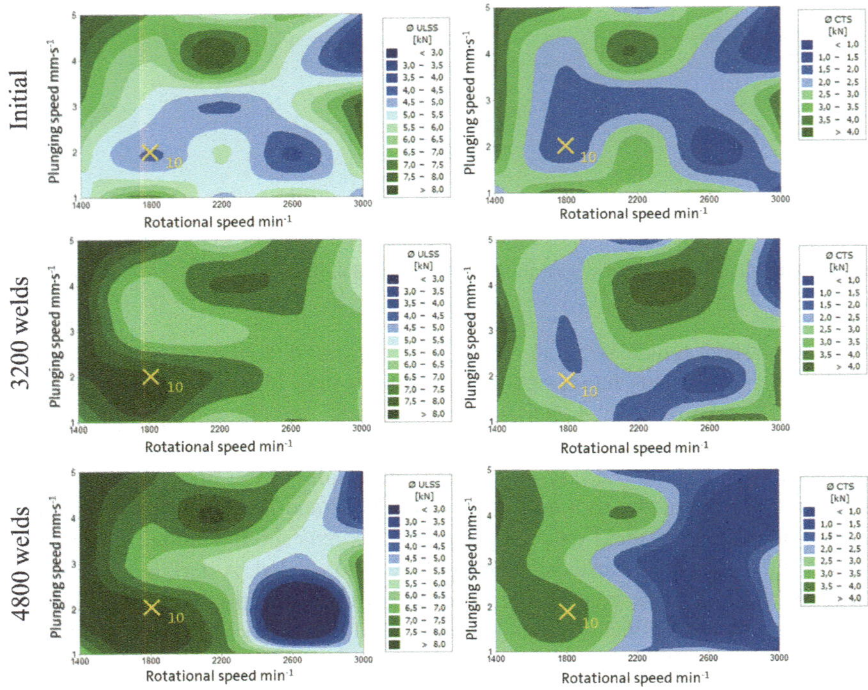

Fig. 5 ULSS and CTS results of the parameter studies with the CrVN tool after tool life tests

are already evident, which are indicative of the respective results in boths lap shear and cross-tensile test. In the initial state, the spot weld shows a refilling defect. The surface of the spot is not smooth or even. The joint between the stirred zone and the surrounding cover plate is interrupted with a gap and stronger air inclusions can be seen at the outer corner in the boundary area of the two plates. Below the stirred zone, the thermomechanically affected zone (TMAZ) can be seen extending semi-circularly downward. At the edges of the weld spot, the depth is not pronounced. After 3200 points, the spot surface is smooth. The air inclusions in the parting line are significantly reduced, which indicates improved shear tensile strength. However, a gap can still be seen between the surrounding top sheet and the stirred zone, which limits the cross tensile strength. The TMAZ goes down steeply at the edges and is limited almost horizontally thereafter. Finally, no defects can be seen in the final micrograph after 4800 spot welds. Cross and shear tensile strength are both in a good range and are on a high level in comparison. Here, too, a horizontal parting line between the TMAZ and base material can be found.

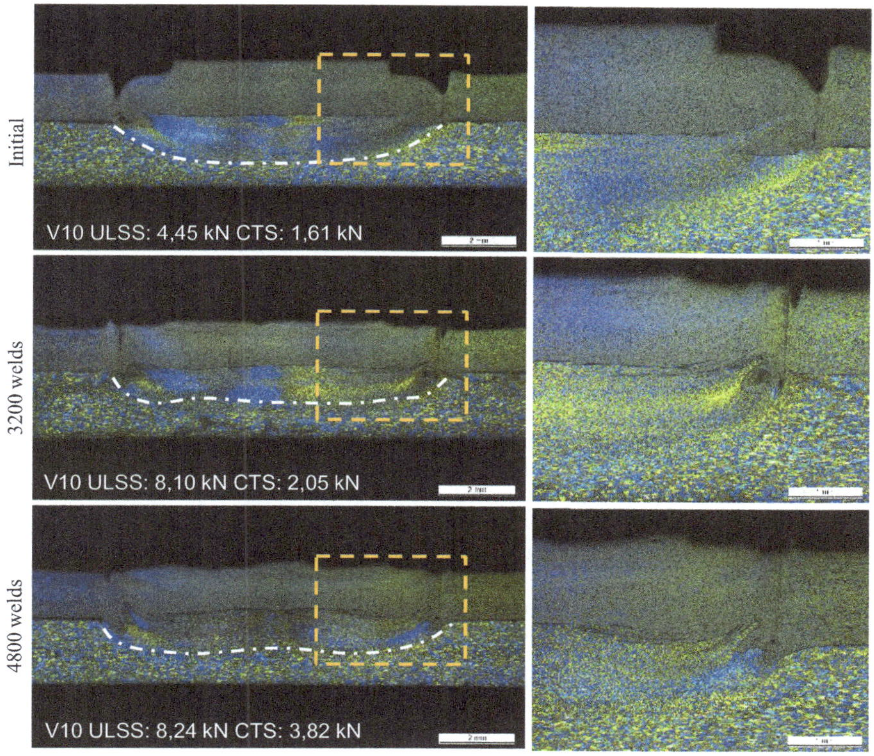

Fig. 6 Cross-section of experiment 10 at the initial state of the tool and after 3200 and 4800 welds

4 Conclusions

The present study could show the influence of the tool properties on the welding result during RFSSW of dissimilar aluminium welds. The results can be concluded as:

- Both coatings, CrVN and Cr-DLC reduced the tool wear, helping the tools reach higher amounts of spot welds than conventional, uncoated tools. For this study, the tool life tests up to 4800 spot welds with each tool are taken into account. At this point, both tools were still intact, the CrVN tool failed after 6027 welds during the tool life tests. The Cr-DLC tool reached 6400 spot welds and remained intact.
- Parameter studies were carried out, leading to ultimate lap shear strengths of up to 9.27 kN and cross-tension strengths reaching up to 5.72 kN with the CrVN coating. The Cr-DLC coated tool reached 9.22 kN and 5.32 kN accordingly. These values were reached, after both tools were at 3200 spot welds during the tool life

tests. The parameter studies in the initial state of the tools lead to lower results with more severe defects.
- The CrVN coated tool broke after 6027 welds during the tool life tests, as the wear had progressed to a severe level. The Cr-DLC coating lead to a better wear protection, resulting in an intact tool after 6400 spot welds. However, the significant temperature exposure during the welding tests resulted in a severe material loss on the outside of the sleeve.
- The welding quality of the Cr-DLC coated tool showed only minor changes. Over the total amount of 6400 spot welds, sound joints could be manufactured. In contrast, the CrVN coated tool showed a significant reduction of the welding quality, which was reflected in the cross-sections as well as in the mechanical results of the tests.

References

Amancio-Filho, S.T., Camillo, A.P.C., Bergmann, L., dos Santos, J.F., Kury, S.E., Machado, N.G.A.: Preliminary investigation of the microstructure and mechanical behaviour of 2024 aluminium alloy friction spot welds. In: The Paper Contains Partial Overlap with the ICAA12 Proceedings by USB under the Permission of the Editorial Committee (2011)

Camilo, A.P., Pieta, G., Brzostek, R., Suhuddin, U., Goushegir, S., Amancio, S., dos Santos, J.F.: A Review on Microstructure and Properties of Fricition Stir Spot Welds (Refill) in Aircraft Aluminium Alloys (2014)

Cao, J.Y., Wang, M., Kong, L., Guo, L.J.: Hook formation and mechanical properties of friction spot welding in alloy 6061–T6. J. Mater. Process. Technol. **230**, 254–262 (2016)

Castro, C.C. de, Plaine, A.H., Dias, G.P., Alcântara, N.G. de, dos Santos, J.F.: Investigation of geometrical features on mechanical properties of AA2198 refill friction stir spot welds. J. Manuf. Process. **36**, 330–339 (2018)

Chai, P., Wang, Y.: Effect of rotational speed on microstructure and mechanical properties of 2060 aluminum alloy RFSSW joint. Met. Mater. Int. **25**, 1574–1585 (2019)

Ferreira, A.C., Campanelli, L.C., Suhuddin, U.F.H., de Alcântara, N.G., dos Santos, J.F.: Investigation of internal defects and premature fracture of dissimilar refill friction stir spot welds of AA5754 and AA6061. Int. J. Adv. Manuf. Technol. **106**, 3523–3531 (2020)

Fukada, S., Ohashi, R., Fujimoto, M., Okada, H.: Refill friction stir spot welding of dissimilar materials consisting of A6061 and hot dip zinc-coated steel sheets. In: Proceedings of the 1st International Joint Symposium on Joining and Welding, Osaka, Japan, 183–187 (2013)

Goushegir, S.M.: Friction spot joining of metal-composite hybrid structures. Dissertation, Technical University Hamburg-Harburg (2015)

GSI SLV München: Reibpunktschweißen von Überlappverbindungen an Aluminiumknet- und –gusslegierungen im Vergleich, München; Abschlussbericht 5151/12 (2010)

Hovanski, Y., Curtis, A., Michaelis, S., Blackhurst, P., Larsen, B.: Advances in refill spot welding productivity. In: Friction Stir Welding and Processing XI in The Minerals, Metals & Materials Society (2021)

Kubit, A., Kluz, R., Trzepieciński, T., Wydrzyński, D., Bochnowski, W.: Analysis of the mechanical properties and of micrographs of refill friction stir spot welded 7075–T6 aluminium sheets. Arch. Civil Mech. Eng. **18**(1), 235–244 (2018)

Lauterbach, D., Keil, D., Harms, A., Leupold, C., Dilger, K.: Tool wear behaviour and the influence of wear-resistant coatings during refill friction stir spot welding of aluminium alloys. Weld. World. **2**, 234–250 (2021)

Li, Z., Ji, S., Ma, Y., Chai, P., Yue, Y., Gao, S.: Fracture mechanism of refill friction stir spot-welded 2024–T4 aluminum alloy. Int. J. Adv. Manuf. Technol. **86**, 1925–1932 (2016)

Li, G., Zhou, L., Luo, L., Wu, X., Guo, N.: Microstructural evolution and mechanical properties of refill friction stir spot welded alclad 2A12-T4 aluminum alloy. J. Mater. Res. Technol. **8**, 4115–4129 (2019)

Montag, T., Wulfsberg, J.-P., Hameister, H., Marschner, R.: Influence of tool wear on quality criteria for refill friction stir spot welding (RFSSW) Process. Procedia CIRP. **24**, 108–113 (2014)

Nasiri, A.M., Shen, Z., Hou, J.S.C., Gerlich, A.P.: Failure analysis of tool used in refill friction stir spot welding of Al 2099 alloy. Eng. Fail. Anal. **84**, 25–33 (2018)

Paschke, H., Weber, M., Kaestner, P., Braeuer, G.: Influence of different plasma nitriding treatments on the wear and crack behavior of forging tools evaluated by Rockwell indentation and scratch tests. Surf. Coat Technol. **205**(5), 1465–1469 (2010)

Tier, M.D., Rosendo, T.S., Mazzaferro, J.A., Mazzaferro, C.P., dos Santos, J.F., Strohaecker, T.R.: The weld interface for friction spot welded 5052 aluminium alloy. Int. J. Adv. Manuf. Technol. **90**(1–4), 267–276 (2017)

Xu, Z., Li, Z., Ji, S., Zhang, L.: Refill friction stir spot welding of 5083-O aluminum alloy. J. Mater. Sci. Technol. **34**(5), 878–885 (2018)

Effect of Welding Parameters on Mechanical Properties of Dissimilar Friction Stir Lap Welds of AA5052 and AISI 1010

L. N. Tufaro and H. G. Svoboda

Abstract Friction stir lap welding (FSLW) is a solid-state welding process which allows joining of similar and dissimilar materials. Welding of dissimilar materials involving aluminium alloys and steels are a current requirement for several industries, mainly for automotive and transportation. Al/steel joints presents several challenges, due to its different physical properties and the formation of intermetallic compounds. FSLW is a particularly interesting welding technique for these joints, especially for thin sheets. Nevertheless, there is lack of information about its application. The welding parameters affect the microstructural evolution, the material flow, defects level, distortions, etc. The aim of this work is to analyse the effect of the travel speed on the mechanical properties of friction stir lap welds of AA5052 aluminium alloy and AISI 1010 low carbon steel in thin sheets (1 and 0.8 mm, respectively). Dissimilar friction stir lap welds of these materials were produced using a tool made of H13 tool steel. Diverse travel speeds were analysed (98, 146 and 206 mm/min). Macrostructure examination, microhardness profiles and maps were performed in the weld cross section. Lap shear tests were conducted, and the fracture surfaces of tested samples were analysed. The maximum load per width unit increased linearly with travel speed, reaching 190 N/mm and a joint efficiency of 73%, and is controlled by the amount of steel inclusions in the aluminium plate through, which decreases with travel speed.

Keywords Friction stir lap welding (FSLW) · Al to steel dissimilar joints · Travel speed · Hook · Steel inclusions

---

L. N. Tufaro
Centro de Investigación y Desarrollo en Mecánica, Instituto Nacional de Tecnología Industrial, Av. Gral. Paz 5445, B1650WAB San Martín, Argentina
e-mail: ltufaro@inti.gob.ar

L. N. Tufaro · H. G. Svoboda (✉)
INTECIN, Universidad de Buenos Aires, Paseo Colón 850, C1127AAR Buenos Aires, Argentina
e-mail: hsvobod@fi.uba.ar

H. G. Svoboda
CONICET, Godoy Cruz 2290, C1033AAJ Buenos Aires, Argentina

L. F. M. da Silva et al. (eds.), *2nd International Conference on Advanced Joining Processes (AJP 2021)*, Proceedings in Engineering Mechanics,
https://doi.org/10.1007/978-3-030-95463-5_5

1 Introduction

Dissimilar joints are widely used in multi-materials structures and thin sheet overlap Aluminium to Steel joints are especially relevant in automotive industry.

Friction stir lap welding (FSLW) is a solid-state welding process which allows joining similar and dissimilar materials. The capability of FSLW for the joining of dissimilar material, such as aluminium and steel, lies in the low heat input that minimize the extent of intermetallic compound (IMC) formation during the welding process (Kumar et al. 2015; Kar et al. 2020; Sorger et al. 2017).

In friction stir welding (FSW) of steels, the most common tool materials are WC and PCBN (Mishra and Mahoney 2007), although it was reported that tool steel can be used in FSLW of Al to Steel joints when the steel is place in the bottom of the joint and the tool pin plunge in the steel is null or low (Chen et al. 2013; Kumar et al. 2015). The level of interaction between the pin and the steel sheet will affect the tool life as well as the mixing between aluminium and steel. It was reported that a slight pin penetration in the steel plate (P_P) of 0.1 mm has not produced a significant wear of the pin (Chen et al. 2013; Kumar et al. 2015). Other authors also have used some degree of pin plunge for this type of joint (Coelho et al. 2008; Elrefaey et al. 2005; Kar et al. 2020; Kimapong and Watanabe 2005a, b; Movahedi et al. 2011, 2012; Wan and Huang 2017).

The pin plunge into the steel sheet led to the formation of a steel hook within the aluminium alloy. This hook can affect the mechanical properties of the welded joint. Two geometrical effects can be pointed out related to the hook formation. One is related to the reduction in the effective thickness of the aluminium alloy sheet (Sorger et al. 2017). Other authors have reported some beneficial effect related to the mechanical locking produced by the steel flow into the Al sheet (Movahedi et al. 2011, 2012).

In general, it seems that some degree of interaction between the pin and the steel sheet is needed to perform the FSLW of aluminium to steel Nevertheless, Chen et al. (2013) have reported that the optimal condition corresponds with a null P_P but with the pin close to the steel plate in order to produce a continued interface intermetallic layer which is the condition for a continued metallurgical bonding.

The mechanical properties of this type of joints are a key aspect for its application. Lap shear test is usually used to evaluate the strength and integrity of the joint. The fracture mode in this test depends on the aluminium/steel interface integrity and local materials strengths and thicknesses. The fracture can produce through the aluminium/steel interface in a shear mode (Chen et al. 2013, Kar et al. 2020, Kimapong and Watanabe 2005a, b, Movahedi et al. 2011, Movahedi et al. 2012, Wan and Huang 2017), or through the thickness of one of the base materials in a tensile mode (Coelho et al. 2008; Kimapong and Watanabe 2005a; Movahedi et al. 2011, 2012). For the first fracture type, it was found that an increase in travel speed (U) and a decrease in rotational speed (ω) produce a reduction of the heat input, resulting in a thinner IMC layer and in a higher a fracture load (Kimapong and Watanabe 2005a; Wan and Huang 2017). Kimapong and Watanabe reported that a variation in ω can change the fracture mode, obtaining a fracture located in the aluminium for lower ω with higher fracture loads, while the fracture was located at the interface for higher ω, with lower loads. They also reported that an excessively high U produces a joint with incompletely welded interface, reducing the fracture load (Kimapong and Watanabe 2005a, b).

Movahedi et al. found that a variation in U can also change the fracture mode. They obtained joints with cavities at the interface for higher U, resulting in a fracture located in the interface with lower fracture loads, while defect free joints were produced for lower U, presenting a fracture at the steel sheet with higher fracture loads (Movahedi et al. 2011, 2012).

The effect of other welding parameters was also analysed. An increase in P_P, pin diameter and tool tilt angle produce a reduction in fracture load, which is related with the IMC layer thickness at the interface and defect generation (Kimapong and Watanabe 2005a, b). Not only the thickness of the IMC layer affects the fracture load, but it was also found that grain refinement of the IMC layer highly enhances the joint strength by increasing the fracture toughness of the brittle IMC layer (Beygi et al. 2021).

Coelho et al. (2008) studied FSLW of AA6181-T4 alloy and HC340LA, obtaining a fracture through the aluminium sheet thickness, suggesting that the amount and the size of the steel inclusions in the Al sheet apparently determine the mechanical properties of the joint.

Despite that FSLW of dissimilar joints of aluminium and steel was subject of different works in the last years, there is still a lack of comprehension of the effect of welding conditions and tool geometries on the mechanical behaviour of the joints, especially for thin sheets.

The objectives of this work were to develop a welding procedure for dissimilar AA5052-H32 aluminium alloy and AISI 1010 carbon steel by FSLW in thin sheets and to analyse the effect of the travel speed on the mechanical strength of friction stir lap welds of AA5052 and AISI 1010.

2 Materials and Methods

Dissimilar lap welds of AA5052-H32 and AISI 1010 in thin sheets of 1 mm and 0.8 mm thickness (t), respectively, were produced by FSLW. Table 1 presents the mechanical properties of base materials used in this work.

Table 1 Mechanical properties of base materials

Material	HV (0.3 kg)	$\sigma_{0,2}$ (MPa)	σ_{UTS} (MPa)	e (%)
AA5052-H32	65	205	260	13.5
AISI 1010	95[a]	180[a]	320[a]	28[a]

[a]Minimum values

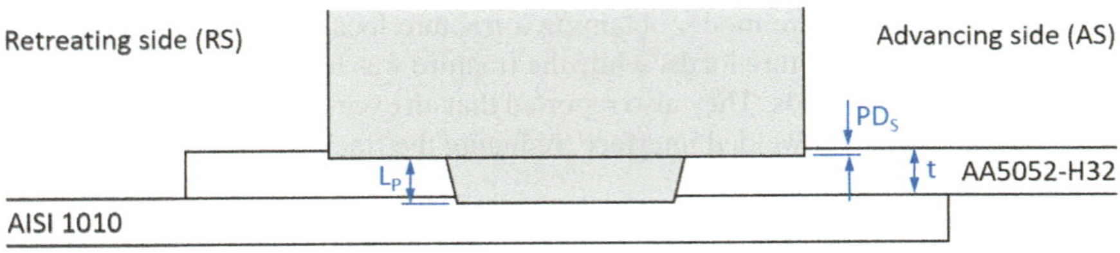

Fig. 1 Scheme of FSLW process

The test coupons were formed by samples of each sheet of 50 mm width and 125 mm long, overlapped 25 mm, as it is shown in Fig. 1. AA5052-H32 was located as the upper sheet, in the advancing side (AS), while AISI 1010 steel sheet was located as the lower one, in the retreating side (RS).

The tool was made of H13 tool steel with a concave shoulder of 12 mm in diameter and a tapered pin, with major minor diameters of 6.2 and 5.7 mm, respectively. The pin length (L_P) was 1 mm like the aluminium alloy plate t, in order to obtain a pin penetration (P_P) in the steel plate similar to the shoulder plunge depth (PD_S). The rotational speed was 680 rpm and different travel speeds (U) were analysed: 98, 146 and 206 mm/min, using this parameter for the sample identification. The tilt angle was 1.5° and the shoulder plunge depth was of the nominally of 0.1 mm.

Cross sections of welded samples were prepared for metallographic observation. Macrostructure examination was performed in order to analyse the material mixing at the aluminium/steel interface and for the measurement of dimensional characteristics of the joints (Fig. 2): welding nugget thickness corresponding to the Al sheet (t_{WN}) and hook position from the weld centreline (P_{Hook}).

The resulting underfill (h) was calculated as the difference between the WN and the parent material (PM) thicknesses ($h = t - t_{WN}$). Dimensional aspects specified on ISO 25392 were verify in order to analyse the qualification of the welding procedure (ISO 25239-4: 2011).

Vickers microhardness profiles (HV) were carry out along the mid-thickness line of the aluminium alloy plate with a spacing of 0.60 mm. Also, Vickers microhardness maps ($HV_{0.1}$) were performed in the aluminium plate at the hook region. The maps were centred in the mid thickness of the PM, with a horizontal and vertical spacing

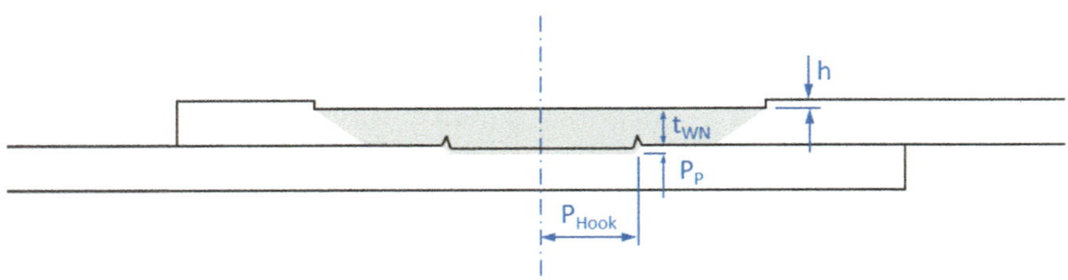

Fig. 2 Scheme of joint dimensional characteristics

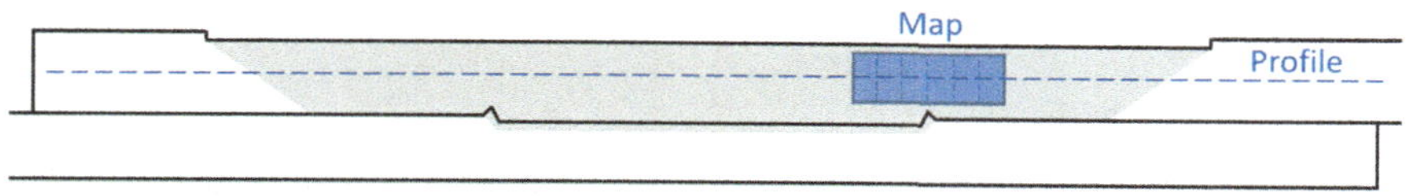

Fig. 3 Scheme of Vickers microhardness measurement locations for profiles and maps

between indentations of 0.30 mm. A scheme of the Vickers microhardness profile and map locations is shown in Fig. 3.

Two specimens from each welding condition were extracted to perform lap shear tests. Load–displacement curves were recorded obtaining the maximum load per width unit (P_{max}/w) for each case. On tested samples of each condition, longitudinal sections were prepared to microstructural examination in order to analyse the fracture location and path. Also, the fracture surface of the specimens was analysed with scanning electron microscopy (SEM) and energy dispersive X-ray spectrometry (EDS).

3 Results and Discussions

Figure 4 shows the macrographs of the weld cross section for the different welding conditions. For all cases, interaction between the aluminium and the steel was obtained. It can be seen a good aluminium/steel interface appearance without showing voids or other macrographic defects.

Geometric characteristics determined are shown in Table 2. The t_{WN} was of the order of 0.9 mm, related with an h of 0.1 mm, although for the 146 mm/min condition the t_{WN} was slightly lower. Despite of that, all the joints satisfy the acceptance limit for the underfill of the ISO 25239, which is 0.15 mm for 1 mm thickness plates (ISO 25239-4: 2011). The P_{Hook} and the extension of the aluminium/steel interface were decreased with U. The P_{Hook} is indicated with an arrow in the macrographs (Fig. 4).

Steel inclusions (SI) are observed embedded into the aluminium plate, mainly located at the hook region. The amount of SI also decreased with U. Even though the pin penetration in the steel plate has a significant impact in the material flow at the interface, the slightly variations in the pin penetration (related with the underfill) do not explain the observed changes in the material flow. In this sense, the changes in material flow can be related with the U.

From the macroscopic examination it could be said that the developed procedure showed to be adequate to achieve a suitable material mixing and could be qualify under ISO 25239, considering this standard is developed for aluminium alloys (ISO 25239–4: 2011).

Vickers microhardness profiles in the AA5052 for the three welding conditions are shown in Fig. 5 with a schematic of a weld cross section to associate the corresponding position into the weld joint.

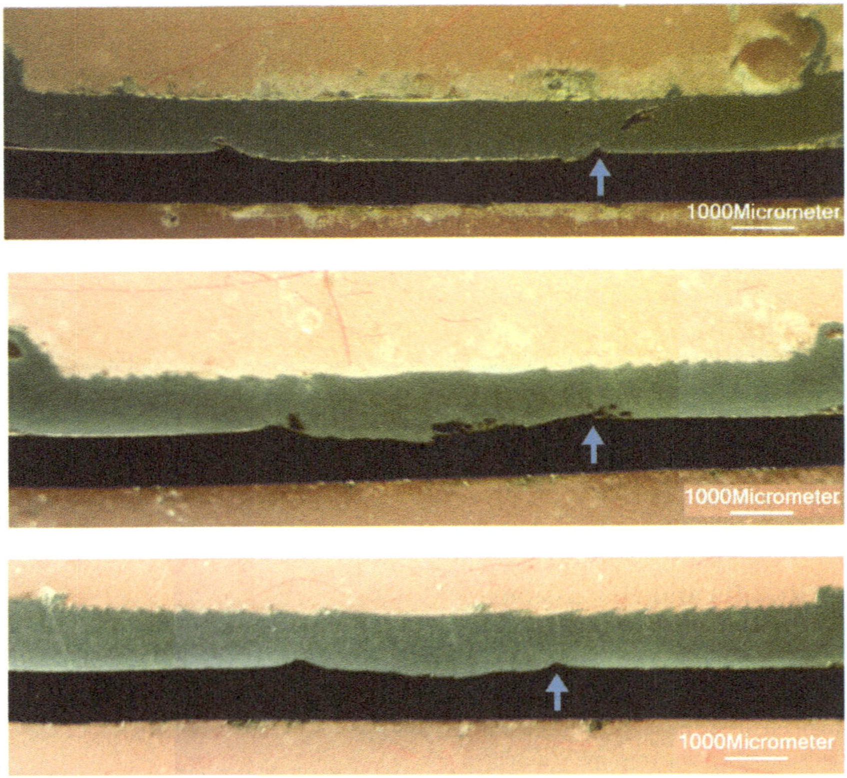

Fig. 4 Macrographs of weld cross section: a. 98, b. 146, c. 206

Table 2 WN geometric characteristics

U (mm/min)	t_{WN} (mm)	h (mm)	P_{Hook} (mm)
98	0.89	0.11	3.06
146	0.86	0.14	2.61
206	0.91	0.09	2.05

The AA5052 is a work hardened aluminium alloy with an average hardness of 75 $HV_{0.3}$. In the HAZ it can be seen a reduction in the hardness, which is related to the recovery and recrystallization due the thermal cycle experimented during the welding process. This behaviour has been reported for work hardened aluminium alloys welded by FSW (Mishra and Mahoney 2007). The hardness in the WN is relatively uniform with an average value of the order of 65 $HV_{0.3}$, similar to the obtained at the high temperature HAZ. Nevertheless, for the lower U (98 mm/min), local values over the average WN hardness were measured, which were associated

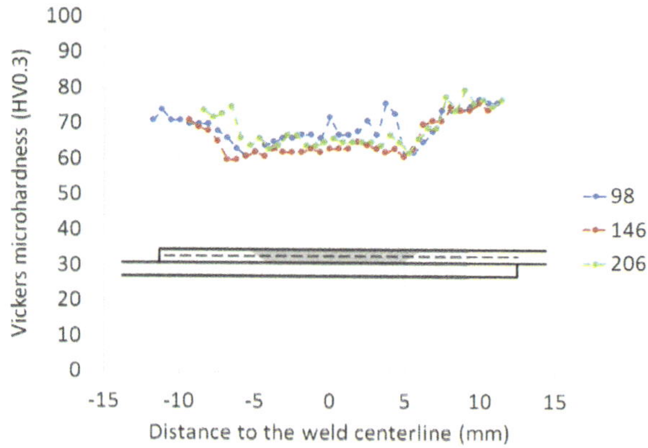

Fig. 5 Vickers microhardness profiles

to the SI in the Al sheet. In general, there is no significant variations of the hardness profiles with U.

Vickers microhardness maps performed in the AA5052 at the hook region for the three welding conditions, with a black and white image of the macrograph displaying the SI distribution in the aluminium alloy sheet, are shown in Fig. 6. The nodes of

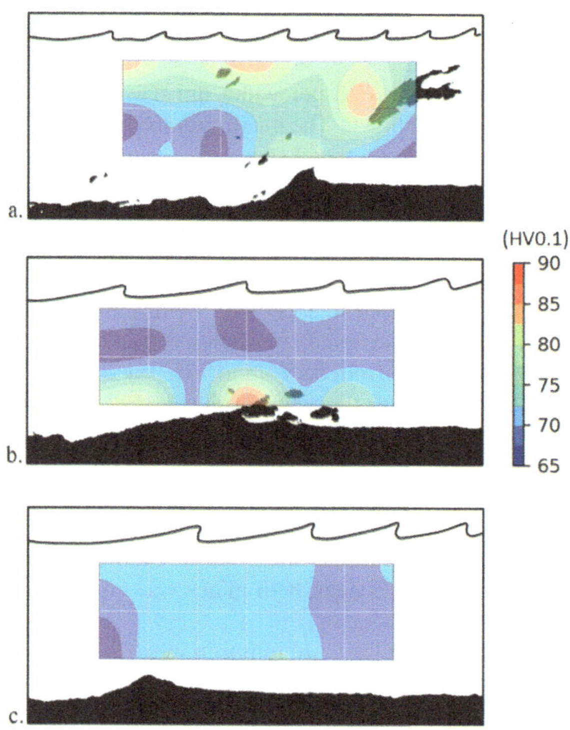

Fig. 6 Vickers microhardness maps ($HV_{0.1}$): **a** 98, **b** 146, **c** 206

Fig. 7 Curves Load per width unit (P/w) - Displacement (Δl) of lap shear tests for different U

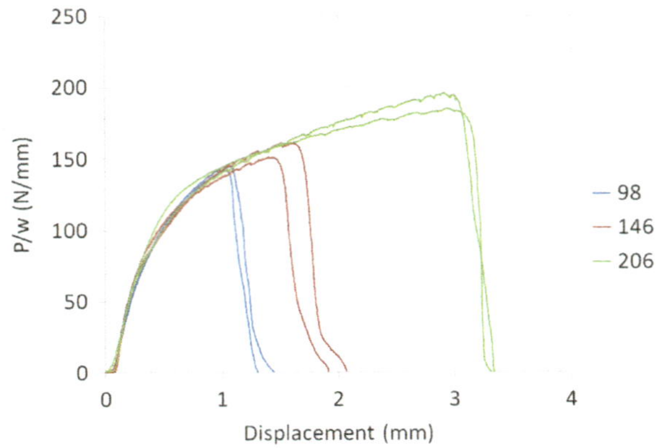

the grid of the maps are associated with the measurement positions.

The hardness maps show values from 65 to 85 $HV_{0.1}$, approximately corresponding to the Al alloy. The uniformity of the hardness in the analysed zone increased with U. The hardness peaks measured are related with the zones adjacent to the SI. In this sense, although thermal softening of the Al alloy could increase as U decreases, the presence of higher amounts of SI could explain the increase in hardness at the hook zone for lower U. For higher U the amount and size of SI decrease, explaining the higher uniformity of the hardness measured and the lower values obtained.

In Fig. 7 the Load per width unit-Displacement curves obtained from the lap shear tests are shown. The maximum load as well as the displacement at maximum load increased with travel speed. Consequently, the total area under the curve also increased with U. It can be seen the low dispersion between both test samples for each condition.

The plot of P_{max}/w vs. U as well as the equivalent joint efficiency (Ef), considering the tensile strength of the AA 5052-H32, is shown in Fig. 8.

P_{max}/w increased linearly with U, reaching an average value of 190 N/mm (for 206 mm/min). The linear correlation obtained shows a good agreement ($R^2 = 0.96$). The Ef increased from 55% up to 73% for 206 mm/min. This maximum Ef reached resulted higher than the reported ones for FSLW of dissimilar joints of work hardened aluminium alloy and steel (Sorger et al. 2017), representing a very promising result.

P_{max}/w, fracture location and mode are defined by the aluminium/steel interface integrity, local materials strengths (microhardness) and joint dimensions (local thicknesses). In all the cases, the fracture was located in the WN of the aluminium plate, presenting a tensile fracture mode, as can be seen in Fig. 9. Therefore, the strength of the interface between both plates was higher than de strength of the aluminium plate at the WN.

P_{max}/w for the U of 146 mm/min was below the obtained linear tendency showed in Fig. 8. This fact could be related with the slightly lower measured t_{WN} for this

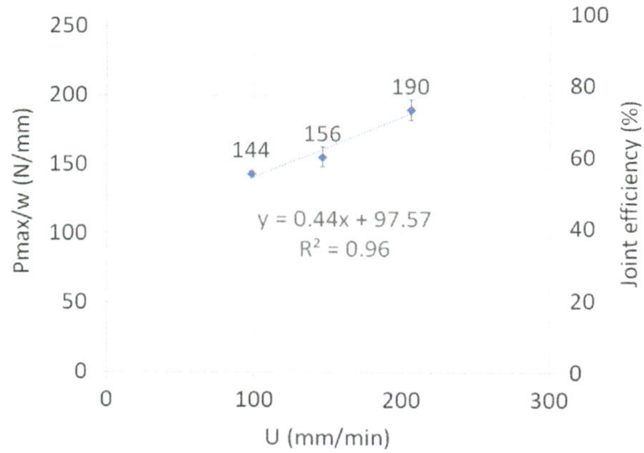

Fig. 8 Maximum load per width unit (P_{max}/w) and joint efficiency (Ef) versus travel speed (U)

Fig. 9 Fractured lap shear test specimens: **a** 98, **b** 146, **c** 206

condition (Table 2). Despite of that, there is no significant effect of the t_{WN} of the AA5052 plate.

The hardness in the fracture location does not exhibit major variations with U, regardless the effect of the SI presence in the aluminium plate. Therefore, the P_{max}/w improvement with increasing U would be controlled by another phenomenon. In Fig. 10 are shown the macro and micrographs of lap shear test specimens, with and without etching.

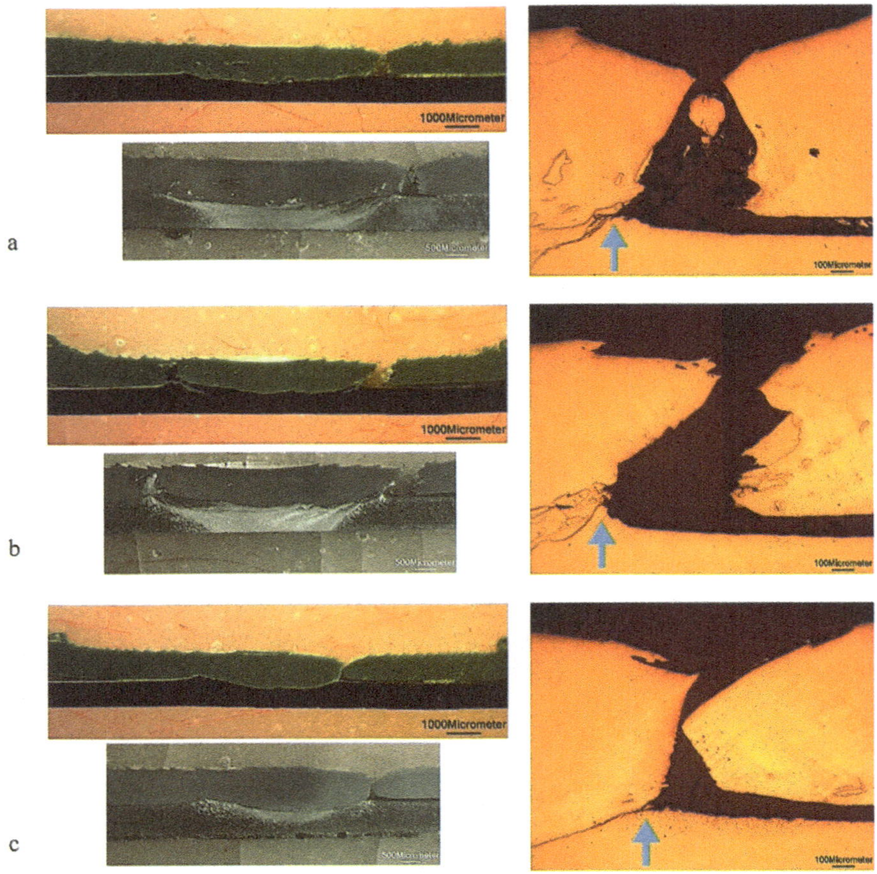

Fig. 10 Macro and micrographs of lap shear test specimens: **a** 98, **b** 146, **c** 206

For all welding conditions, the fracture initiates at the hook position (P_{Hook}), which is indicated with an arrow in the micrographs. It was reported previously the hook reduces locally the t_{WN} of the aluminium plate, promoting the fracture in this position (Sorger et al. 2017). Additionally, the fracture path is associated to the SI in the aluminium plate, which are located mainly in the hook region. As can be seen in Fig. 9a, b, the fracture propagated partially through the aluminium/SI interface, probably due to the presence of the brittle intermetallic compound (IMC) (Chen et al. 2013; Sorger et al. 2017).

For the higher U, lower amounts of SI were in the fracture path (Fig. 9c). Therefore, P_{max}/w would be controlled by the amount of SI in the aluminium sheet at the hook region in the t_{WN}, which decreased with the increase of U, producing an improvement in P_{max}/w. This lower amount of SI is associated to the lower interaction between the tool pin and the steel sheet, which is also affected by h. For this reason, the sample of 146 mm/min, which has a higher h, showed a higher interaction and steel flow into Al sheet, with a consequently lower fracture load.

Related to the Ef, it must be considered that the thinning produced et al. sheet in the WN due to h and also by the hook affect the maximum value of Ef that it could be reached. This is especially relevant in thin sheets.

Figure 11 shows SEM images and EDS maps of Al and Fe of the fracture surface of the fractured lap shear test specimen for 146 mm/min.

The fracture surface shows two zones with different appearance (Fig. 11a). The upper zone presented a ductile fracture showing dimples with plastic deformation in shear (Fig. 11b) and only aluminium was detected on it (Fig. 11d). The lower zone presented a more brittle fracture (Fig. 11c) and both Fe and Al detected on it (Fig. 11e). The presence of both elements confirms that this zone corresponds to a SI.

This observation is consistent with what was observed previously, associated to the fracture path, confirming it follows the SI. Probably, the brittle fracture surface is associated to the IMC layer in the aluminium/steel interface between the aluminium matrix and the SI. Similar fractographies were obtained by Chen et al. (2013) for FSLW of dissimilar joints of aluminium alloy and steel (Chen et al. 2013).

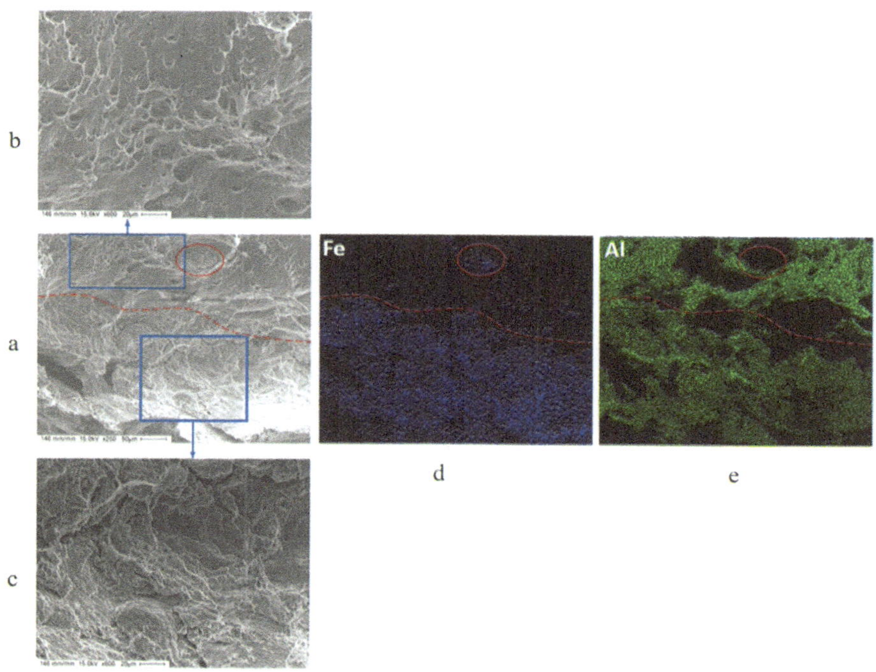

Fig. 11 SEM images and EDS mapping of Al and Fe of the fracture surface of the fractured lap shear test specimen for 146 mm/min

4 Conclusions

The effect of travel speed on mechanical properties and failure mechanism in dissimilar friction stir lap welding of AA5052-H32 and AISI 1010 in thin plates of 1 mm and 0.8 mm, respectively, was studied. The developed procedure showed to be adequate and could be qualify under ISO 25239, considering the scope of this standard.

Interaction between aluminium and steel was observed in the welding nugget, with the presence of steel inclusions in the aluminium welding nugget, especially in the hook zone, which decreased with travel speed. These steel inclusions increased the hardness of the AA5052 in the hook zone.

In lap shear test, interface showed a good integrity for all welding conditions. A tensile fracture mechanism was observed in all cases and the fracture was located at the P_{Hook}. Maximum load per unit width increased linearly with travel speed, reaching 73% of the tensile strength of the AA5052-H32 (for 206 mm/min).

On the fracture surface of tested samples there were identify two zones, one with a ductile fracture associated to the Al matrix, and the second with presence of Al and Fe, associated to the SI.

Steel inclusions in the welding nugget of the AA5052 controlled the fracture location, path and the maximum load per unit width.

To optimize the fracture load in this type of joints, the interaction between Al and Steel has to be high enough to produce an interface with good integrity but low enough to minimize the steel inclusions into the Al sheet thickness. This balance will lead to reach the maximum joint efficiency.

Acknowledgements The authors acknowledge to Universidad de Buenos Aires and Agencia Nacional de Promoción Científica y Tecnológica (ANPCYT) for the funding of the project (UBACYT 2018 20020170100038BA /PICT 2017 N°3782) and also to the personnel of Instituto de Tecnologías y Ciencias de la Ingeniería and Centro de Investigación y Desarrollo en Mecánica for its collaboration.

References

Beygi, R., Carbas, R.J.C., Barbosa, A.Q., Marques, E.A.S., da Silva, L.F.M.: A comprehensive analysis of a pseudo-brittle fracture at the interface of intermetallic of η and steel in aluminum/steel joints made by FSW: microstructure and fracture behavior. Mater. Sci. and Eng. A **824**, 141812 (2021). https://doi.org/10.1016/j.msea.2021.141812

Chen, Z.W., Yazdanian, S., Littlefair, G.: Effects of tool positioning on joint interface microstructure and fracture strength of friction stir lap Al-to-steel welds. J. Mater. Sci. **48**(6), 2624–2634 (2013). https://doi.org/10.1007/s10853-012-7056-0

Coelho, R.S., Kostka, A., Sheikhi, S., Dos Santos, J., Pyzalla, A.R.: Microstructure and mechanical properties of an AA6181-T4 aluminium alloy to HC340LA high strength steel friction stir overlap weld. Adv. Eng. Mater. **10**(10), 961–972 (2008). https://doi.org/10.1002/adem.200800028

Elrefaey, A., Gouda, M., Takahashi, M., Ikeuchi, K.: Characterization of aluminum/steel lap joint by friction stir welding. J. Mater. Eng. Perform. **14**(1), 10–17 (2005). https://doi.org/10.1361/10599490522310

International Organization for Standardization: Friction stir welding—aluminium—part 4: specification and qualification of welding procedures (ISO 25239-4:2011). ISO (2011). https://www.iso.org/

Kar, A., Vicharapu, B., Morisada, Y., Fujii, H.: Elucidation of interfacial microstructure and properties in friction stir lap welding of aluminium alloy and mild steel. Mater. Charact. 168, 110572 (2020). https://doi.org/10.1016/j.matchar.2020.110572

Kimapong, K., Watanabe, T.: Lap joint of A5083 aluminum alloy and SS400 steel by friction stir welding. Mater. Trans. JIM **46**(4), 835–841 (2005a). https://doi.org/10.2320/matertrans.46.835

Kimapong, K., Watanabe, T.: Effect of welding process parameters on mechanical property of FSW lap joint between aluminum alloy and steel. Mater. Trans. JIM **46**(10), 2211–2217 (2005b). https://doi.org/10.2320/matertrans.46.2211

Kumar, N., Yuan, W., Mishra, R.S.: Friction Stir Welding of Dissimilar Alloys and Materials. Butterworth-Heinemann, Oxford (2015)

Mishra, R.S., Mahoney, M.W. (ed.): Friction Stir Welding and Processing. ASM International, Ohio (2007)

Movahedi, M., Kokabi, A.H., Reihani, S.S., Najafi, H.: Mechanical and microstructural characterization of Al-5083/St-12 lap joints made by friction stir welding. J. Proc. Eng. **10**, 3297–3303 (2011). https://doi.org/10.1016/j.proeng.2011.04.544

Movahedi, M., Kokabi, A.H., Reihani, S.S., Najafi, H.: Effect of tool travel and rotation speeds on weld zone defects and joint strength of aluminium steel lap joints made by friction stir welding. Sci. Technol. Weld. Joining **17**(2), 162–167 (2012). https://doi.org/10.1179/1362171811Y.0000000092

Sorger, G., Wang, H., Vilaça, P., Santos, T.G.: FSW of aluminum AA5754 to steel DX54 with innovative overlap joint. Weld World **61**(2), 257–268 (2017). https://doi.org/10.1007/s40194-016-0412-y

Wan, L., Huang, Y.: Microstructure and mechanical properties of Al/Steel friction stir lap weld. Metals **7**(12), 542 (2017). https://doi.org/10.3390/met7120542

Nickel-Iron-Alloy Modification to Enhance Additively Welded Microstructure for Subsequent Milling

A. Eissel, L. Engelking, K. Treutler, D. Schroepfer, V. Wesling, and T. Kannengiesser

Abstract The aerospace industry uses nickel–iron alloys, e.g., FeNi36, to create moulding tools for composite materials, since these alloys have a low coefficient of thermal expansion. Nickel–iron alloys are hard-to-cut materials. The moulding tools are large in size and involve complex structures, making them cost-intensive and difficult to manufacture. Thus, the focus is set on additive manufacturing, which can additionally enable the repair of components in order to eliminate local defects. However, the process usually results in a heterogeneous microstructure and anisotropic mechanical properties. As there is a high demand for a precise and exact fit of the precision moulds and the surface quality, the welded components must be subsequently machined. Additionally, inhomogeneous microstructure may lead to unstable cutting forces and conditions. Consequently, a modification of the microstructure morphology is achieved through specific alloy modifications in order to stabilise and improve the subsequent machining process. Therefore, titanium and zirconium are chosen as modification elements with a maximum 1% weight percent and are added to nickel–iron alloy powder. The elements are alloyed, and build-up

A. Eissel (✉) · K. Treutler · V. Wesling
Institute of Welding and Machining, University of Technology Clausthal, Agricolastraße 2, 38678 Clausthal-Zellerfeld, Germany
e-mail: antonia.eissel@tu-clausthal.de

K. Treutler
e-mail: kai.treutler@tu-clausthal.de

V. Wesling
e-mail: volker.wesling@tu-clausthal.de

L. Engelking · D. Schroepfer · T. Kannengiesser
Federal Institute for Material Research and Testing (BAM), Unter den Eichen 87, 12205 Berlin, Germany
e-mail: lorenz.engelking@bam.de

D. Schroepfer
e-mail: dirk.schroepfer@bam.de

T. Kannengiesser
e-mail: thomas.kannengiesser@bam.de

L. F. M. da Silva et al. (eds.), *2nd International Conference on Advanced Joining Processes (AJP 2021)*, Proceedings in Engineering Mechanics,
https://doi.org/10.1007/978-3-030-95463-5_6

welded by plasma-transferred-arc welding. The resulting microstructure morphology of the welded wall structure and the machining properties are then determined. It can be shown that titanium has a significant effect on the structural morphology of the welded layers, as well as on the machining.

Keywords Modification of structural morphology · FeNi36 (Invar) · Plasma transferred arc welding · Ultrasonic-assisted milling

1 Introduction

Since the Swiss physicist Charles Édouard Guillaume developed the FeNi36 alloy in 1897, this alloy has become a major field of research. In particular the aerospace industry pushed it in the mid-1980s, as nickel–iron alloys were used for molding tools for composites (Sahoo and Medicherla 2021; Salgueiriño-Maceira et al. 2006). Fe–Ni alloys with a nickel concentration of 36% and a face-centred cubic structure have a remarkably low thermal expansion coefficient of 1.7–2.0 × 10^{-6} K^{-1}, even at higher temperatures—known as the invar effect (Salgueiriño-Maceira et al. 2006; Berns and Theisen 2008; Lagarec et al. 2001)—and are therefore ideal such applications. The required laminating tools made of FeNi36 for composites production can be up to 12 m in length and are quiet heavy in weight. The focus is on length and size stabilization even at the higher temperatures in the autoclave moulding. On the one hand, the aim is to build more complex structures and, on the other hand, to repair the components in order to eliminate local defects. For this purpose, additive manufacturing is especially useful. Additively manufactured components exhibit a heterogeneous microstructure and anisotropic mechanical properties (Guévenoux et al. 2020). Nickel–iron alloys are difficult to machine (Zheng et al. 2015) and an inhomogeneous microstructure also leads to unstable cutting forces. The manufactured components additionally require subsequent surface finishing because of precise aerodynamic surfaces and an exact fit and thus a high surface integrity. In general, metals are machined with geometrically defined cutting edges, i.e. turning and milling (Schroepfer et al. 2021). Surface integrity is a key requirement for safety-relevant components and is a complex interaction between topographic (surface defects), metallurgical (microstructural alterations) and mechanical (residual stresses) effects (Ulutan and Ozel 2011). With the use of hybrid machining processes, such as ultrasonic-assisted milling (USAM), a significant improvement in surface integrity can be achieved. In the USAM process, the conventional milling process is superimposed with a high-frequency vibration. Rinck et al. achieved for the USAM of Ti–6Al–4V besides a significant reduction of the cutting forces an increased surface integrity compared to the conventional milling process (Rinck et al. 2020). Ni et al. also reported a reduction in the cutting forces of the USAM compared to the conventional milling process for the same material. In addition, the authors found a significant improvement in machining defects and a reduction in surface roughness (Ni et al. 2018).

Within a research project (IGF No. 20.979N) an approach is investigated, with which the microstructure morphology by specific alloy modifications is influenced in order to stabilize and improve the subsequent machining process. Studies on the refinement of FeNi36 alloys on vacuum arc melting furnaces are used as a starting point (Chen et al. 2019; Sui et al. 2019; Yu et al. 2013; Abbasi et al. 2015). Therefore, titanium and zirconium were selected as modification elements with maximum 1 wt.-% and are added to nickel–iron base alloy. These elements have shown a refinement of the morphology structure in recent studies of other superalloys, e.g., Co-Cr alloy (Eissel et al. 2021). The elements are alloyed, and build-up welded by using plasma-transferred-arc welding. The resulting microstructure morphology of the welded wall structure and the machining properties are then determined. The aim of the investigation is the analysis of potential positive effects due to the altered microstructure for the subsequent conventional and ultrasonic-assisted machining processes.

2 Materials and Methods

2.1 Experimental Setup

FeNi36 initial alloy and the two modifications with 1% titanium and 1% zirconium were mixed and homogenised using a 3D shaking mixer TURBULA® by rotational and translational movements and inversion. The titanium powder was provided by Corodur Fülldraht GmbH. All other powders for the respective elements were obtained from the company ECKART GmbH. Plasma-Transferred-Arc (PTA) was used for the welding experiments, using a welding system by Hettiger Schweißtechnik GmbH. A sandblasted sheet with almost identical chemical composition to FeNi36 was selected as the substrate plate in order to minimize the accumulation of foreign elements. An analysis of the substrate plate was determined using spark spectrometry. The chemical composition of the welding powder used for the initial alloy and modifications, as well as the substrate sheet, is listed in the Table 1.

A wall structure of dimensions (50 (length) × 50 (width) × 18 (height)) mm^3 was welded for the microstructure and hardness measurements as well as for the machining investigations, cf. Fig. 1. In this process, the alloy systems are always

Table 1 Chemical composition of welding powder and substrate sheet in wt.-%

	Fe	Ni	Ti	Zr	Mn	Si
FeNi36	63	36	–	–	–	–
FeNi36 + 1% Ti	63	36	1	–	–	–
FeNi36 + 1% Zr	63	36	–	1	–	–
Substrate	~ 62.63	36.55	–	–	0.39	0.1

Fig. 1 Plasma transferred-arc (left) and FeNi36 wall structure via PTA (right)

Table 2 Welding process influencing variables

	Welding gas	Welding current (A)	Welding voltage (V)	Welding speed (cm/min)	Width of pendulum (mm)	Intermediate layer temperature (°C)
FeNi36	Ar 4.6	180	21	20	18	120
FeNi36 + 1% Ti	Ar 4.6	220	21	20	18	120
FeNi36 + 1% Zr	Ar 4.6	200	21	20	18	120

applied one layer on top of the other in one direction and the specified width is achieved by a pendulum motion of the torch.

The interlayer temperature was 120 °C and all specimens were welded with argon. An argon atmosphere was also applied in the arc and induction melting furnace studies (Chen et al. 2019; Abbasi et al. 2015). The welding current varies depending on the modification and is listed in Table 2, as well as the other welding process influencing variables.

2.2 *Metallurgical Analysis and Hardness Measurements*

For the metallographic analyses and the hardness measurement, a cross-sectional specimen platelet were eroded from the wall structure. The cross-section were divided into three parts, upper, middle and lower layers. The specimens were embedded, ground and polished. Cold adler was used as etchant for the specimen for approx. 5 s. The Leica CRT 6 LED light microscope was used for light microscopic imaging. Afterwards, the grain boundaries were evaluated with the software IMS Client by Imagic Bildverarbeitung AG. Chemical composition was identified by energy dispersive X-ray spectroscopy (EDX) analysis on a Cambridge Series 4 CamScan.

For a more precise microstructural analysis, two hardness measurements were carried out using the DuroScan G5 series hardness testing machine. Hardness measurements with HV 0.2 were made centrally over the entire specimen cross-section. Additionally, five measurements with HV 10 at different locations of the cross-section, four at the edge region of the specimen and one in the centre of the cross-section, were made.

2.3 *Machining Experiments*

Machining experiments without lubrication were carried out in down milling mode on a 5-axis machining center (DMU 65 by DMG MORI) modified for ultrasonic-assisted milling in order to investigate the influence of the modification of the FeNi36 initial alloy on subsequent finishing milling conditions. In USAM, the tool oscillates in the ultrasonic frequency range in the axial direction with an amplitude of $<10\ \mu m$. The milling tool for the experiments was a PVD-coated (TiAlN/TiSiN-based) solid carbide ball end mill with a diameter of 6 mm and four flutes (by WOLF Werkzeugtechnik GmbH). The machinability of the FeNi36 initial alloy and the two modifications with 1% Ti and 1% Zr was investigated. The experimental setup is shown in Fig. 2a. The extraction of machining specimens from the AM walls was accomplished by electrical discharge machining (EDM) in as welded condition with dimensions of approx. 16 mm × 16 mm × 2 mm (length × width × height), cf. Fig. 2b. Table 3 shows the applied milling parameters. The tests were carried out with and without ultrasonic-assistance for each alloy modification and both processes were performed with one milling tool, resulting in six milling parameter and material combinations and two milling tools for the tests.

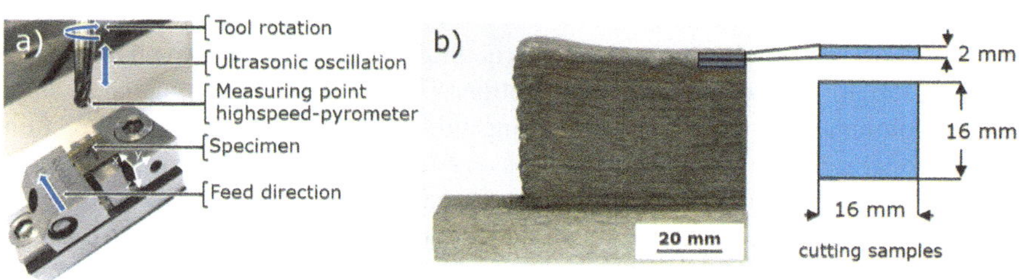

Fig. 2 **a** Experimental setup, **b** the exemplary sampling and the specimen dimensions

Table 3 Experimental data for machining analyses

Cutting speed $v_c = 30$ m/min	Feed rate $f_z = 0.07$ mm/tooth	Ultrasonic assistance $P_{US} = 0\%$ and 100%
Angle in x-direction $\lambda = 45°$	Step over $a_p = 0.3$ mm	Ultrasonic frequency $F_{US} = 38.34$ kHz
Angle in y-direction $\tau = 45°$	Cutting engagement $a_e = 0.3$ mm	Amplitude $\hat{u} = 0.368$ μm

2.3.1 Temperature Analysis

An in-situ measurement of the temperature of the tool cutting (as the cutting edge passes through the air at the reversal point of the tool revolution) was realized using a high-speed pyrometer (KGA 740 by KLEIBER), cf. Fig. 2a.

2.3.2 Surface Analysis

In addition, roughness measurements were carried out using a contact profilometer tester (HOMMEL-ETAMIC T1000 by JENOPTIK) and the arithmetic mean roughness *Ra* was determined in accordance with the DIN EN ISO 4287 standard normal to the feed direction (DIN EN ISO 4287 2010). Furthermore, images of the finish milled surfaces were taken using a light microscope (VHX-7000 by KEYENCE).

3 Results and Discussion

3.1 Microstructure and EDX Analysis

Figure 3 shows the structure of the microstructural morphologies based on the middle layers of the initial alloy and the modifications. The structure is dominated by large vertical columnar grains of austenite growing almost parallel to the build-up direction. A directional solidified structure occurs as a result of rapid solidification.

The PTA weld of the initial FeNi36 alloy exhibits oxides and oxide bands formed in the contact zone between the layers indicated by dark grey particles, see Fig. 4, with diameters between 45 and 80 μm. The oxides are etched out by chemical erosion using adler etchant and can be seen as pores in the image. It is assumed that oxide formation is due to the large surface area of the powder. The welding process then causes these to accumulate at the contact zones.

Cellular grains are developed in the area of the contact zone and the pores. Furthermore, finely distributed smaller pores can be identified in all austenite grains. These

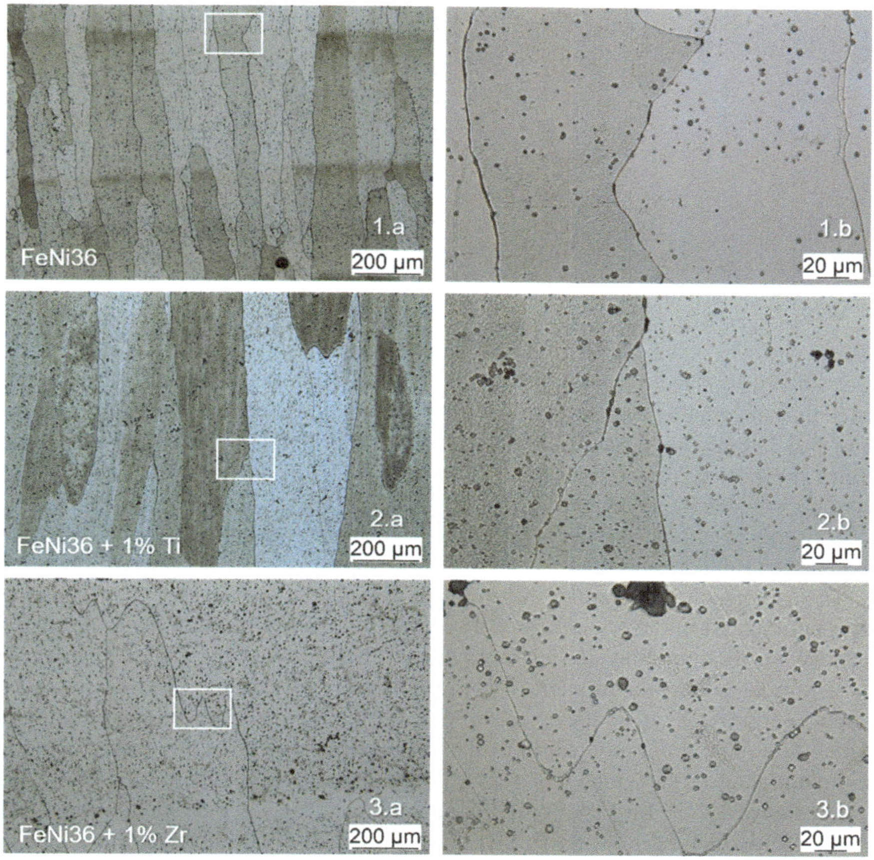

Fig. 3 Microstructure of (1) FeNi36, (2) FeNi36 + 1% Ti, (3) FeNi36 + 1% Zr in 100 × magnification (**a**) and 500 × magnification (**b**)

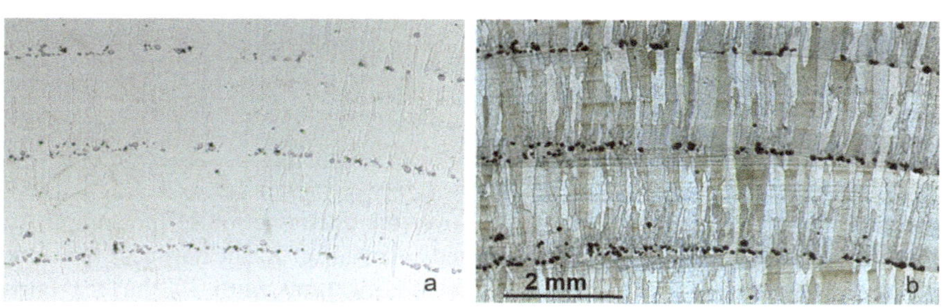

Fig. 4 Oxide bands in contact zone—**a** before etching and **b** after etching

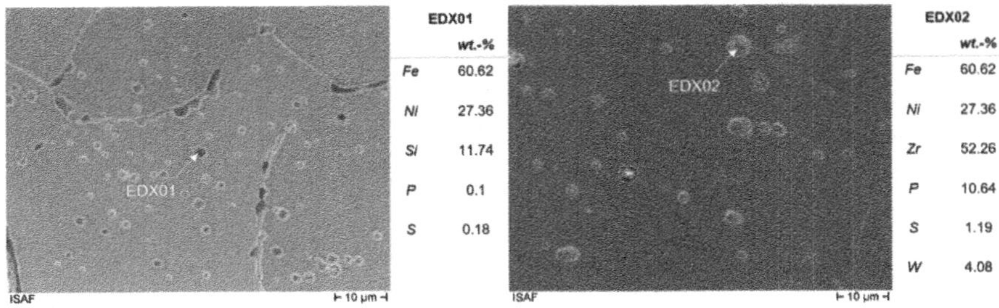

Fig. 5 EDX analysis of occuring precipitations—**a** FeNi36, **b** FeNi36 + 1% Zr

varies between 1 and 4 μm in diameter. The EDX point measurement analysis indicates that the smaller pores are precipitates with silicon, see measurement EDX01 (Fig. 5a). In both modifications of FeNi36, the grains increase in extend, and cellular grains are no longer formed in the contact zone of the layers. Less large oxides develop between the layer, and if they form a confining of the edge areas is obvious. Here further precipitations occur which were formed by the modifying elements, as can be seen in the example of EDX analysis of FeNi36 + 1% Zr (Fig. 5b).

The average chord length of the grains of the initial alloy increases for the upper layers compared to the lower layers. The grain elongation behaves inversely and is reduced for the upper layers in comparison to the lower layers. The modifications show similar behaviour in average chord length. For FeNi36 + 1% Ti the chord length is approx. 200 μm and for FeNi36 + 1% Zr approx. 300 μm longer than the initial alloy. The grain elongation of the modifications behaves contrary compared to the initial alloy, showing an increase for the upper layers of 35.2% for FeNi36 + 1% Ti and of 32.4% for FeNi36 + 1% Zr, cf. Fig. 6. Hence, there is no refinement of the microstructure in the upper layers, but an increase in grain size.

The EDX surface analysis of the upper and middle layers is shown in Table 4. The titanium and zirconium contents are significantly below the powder mixture. It is assumed that large amounts of the modifiers are burning off during the welding process.

3.2 Hardness Measurements

The hardness measurements with HV 0.2 over the cross-section of the sample are shown by means of mathematically smoothed curves, Fig. 7. The hardness measurements of the FeNi36 initial alloy and the modifications with 1% have a fairly steady curve at about the same level. The hardness is between 130 HV 0.2 and 140 HV 0.2. The hardness values match with the data sheet for annealed FeNi36 (Weichmagnetische Eisen-Nickel-Legierungen 2021).

The five individual hardness measurements with HV 10 are summarised in Fig. 8. The average hardness HV 10 of the upper and middle layers is below the initial

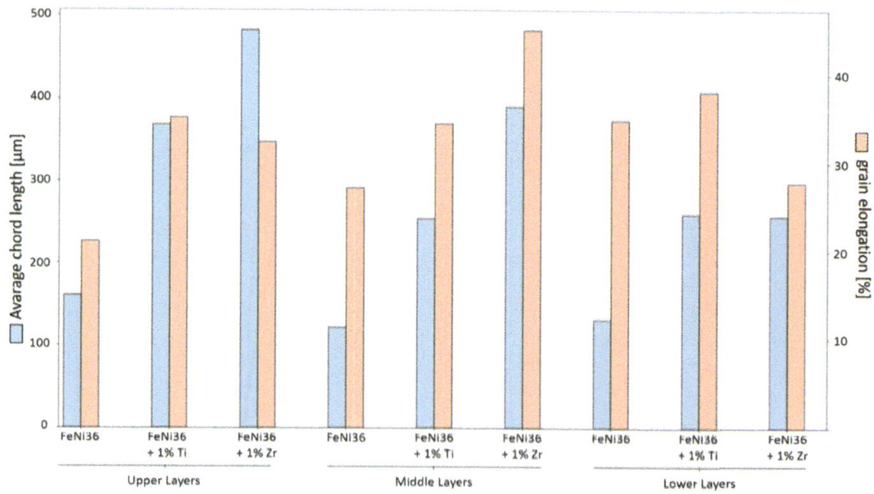

Fig. 6 Average chord length and grain elongation of FeNi36 and modifications

Table 4 EDX surface analysis of upper and middle layers ± error

		Fe	Ni	Ti	Zr
Middle layers	FeNi36	63.93 ± 0.56	35.99 ± 0.62	–	–
	FeNi36 + 1% Ti	65.83 ± 0.69	34.89 ± 0.75	0.21 ± 0.08	–
	FeNi36 + 1% Zr	64.14 ± 0.49	35.15 ± 0.52	–	0.35 ± 0.06

alloy. The average value of the middle layers is 136 HV 10 for FeNi36 + 1% Zr and 131 HV 10 for FeNi36 + 1% Ti. In the lower layers similar hardness levels as in the initial alloy are observable.

3.3 Machining Analysis

3.3.1 Temperature Analysis

The temperature of the tool cutting edge (as the cutting edge passes through the air at the reversal point of the tool revolution) for all machining experiments of the three materials was below the lower limit of the pyrometer's measuring range of 160 °C for. In comparison to analyses for other hard-to-cut materials, e.g., Co-Cr alloy (Engelking et al. 2021), where significantly higher temperatures were observed, it is therefore assumed, that the temperature load of the cutting tool is negligibly small.

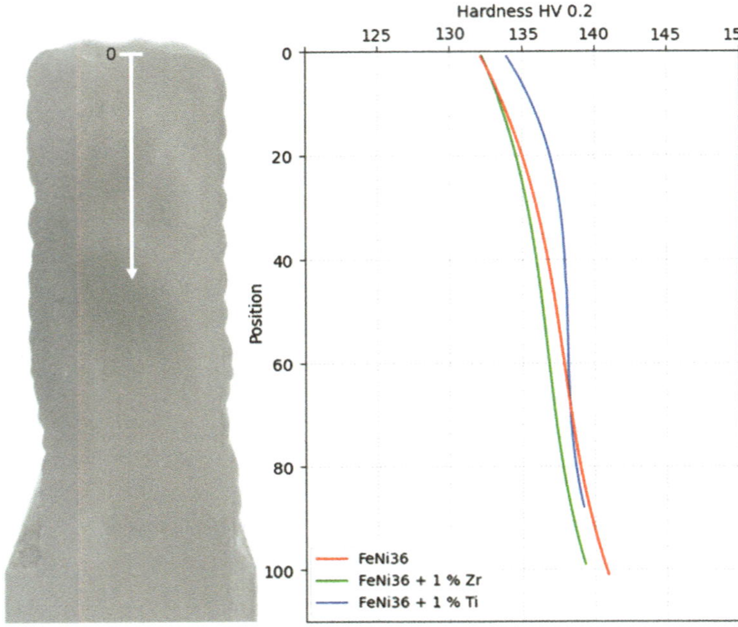

Fig. 7 Hardness measurement series HV 0.2 over the entire cross-section

3.3.2 Surface Integrity Analysis

The results of the roughness measurement are shown in Fig. 9. The USAM causes a lower arithmetic mean roughness *Ra* for all materials. The alloy modifications have no significant influence on the roughness of the surfaces machined via USAM. For the conventional process, the modification of FeNi36 + 1% Ti also has no significant influence on the roughness. The arithmetic mean roughness *Ra* and the error bars of the FeNi36 + 1% Zr are much higher compared to the other two materials. In the course of the roughness measurement, it is assumed that one or more pores within the analysed section affected the measurement, increasing the roughness parameter *Ra* significantly.

Figure 10a–f show pictures of the finish milled surfaces of the FeNi36 initial alloy and both modifications. Figure 10a, c and e exhibit surfaces typical for ball end milling processes (Nespor et al. 2015). Figure 10b, d, and f indicate patterns typical for USAM (Ahmed et al. 2021). The modifications with 1% Ti and 1% Zr cause an increased defect density of the surfaces for the conventional milling process compared to the FeNi36 initial alloy, cf. Fig. 10a, c and e. On the surfaces of the conventional milling process (10a, c and e) as well as on the surfaces of the USAM (Fig. 10b, d and f) local material accumulations can be seen, which are suggested to be a so-called built-up edge (BUE). The low thermal conductivity of FeNi36 combined with its high tendency for work hardening leads to a significant adhesion of the

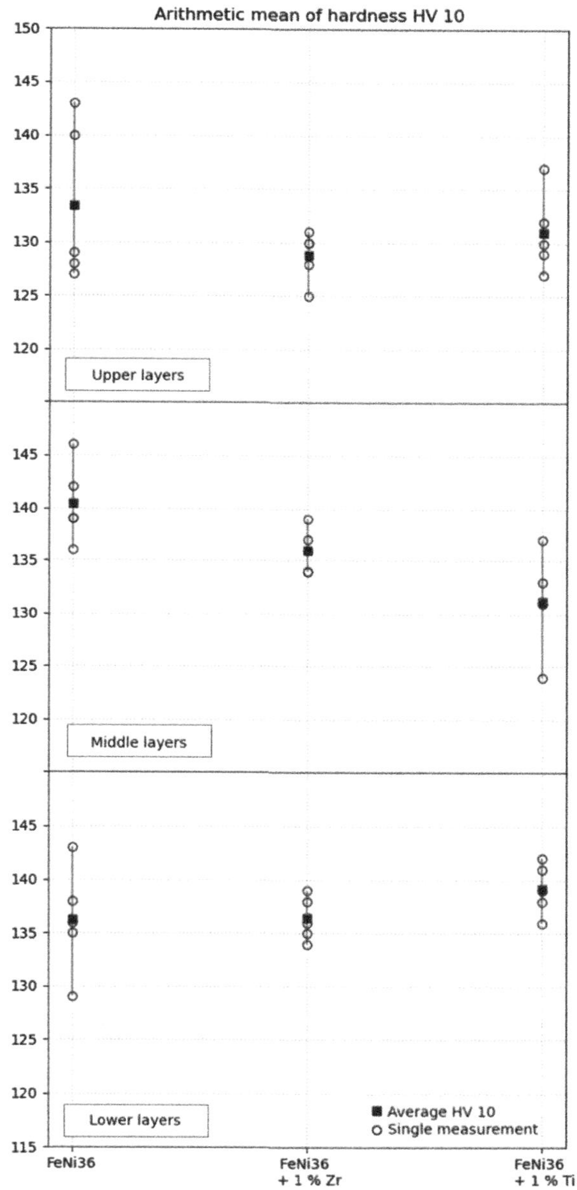

Fig. 8 Hardness measurement HV 10 of FeNi36 initial alloy and the modifications

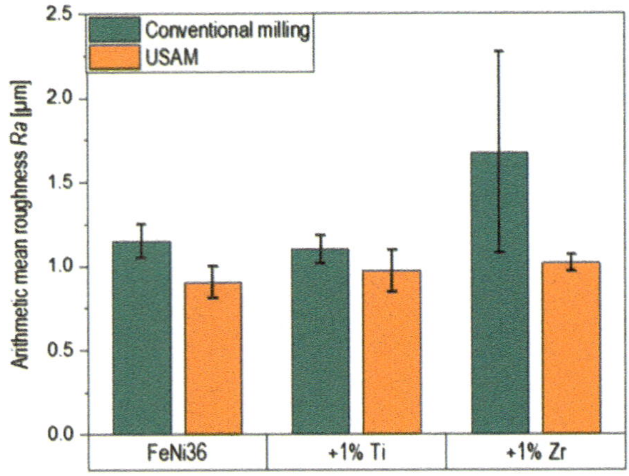

Fig. 9 Arithmetic mean roughness Ra of FeNi36 alloy and the modifications with + 1% Ti and + 1% Zr

material to the tool (Zheng et al. 2015; Ahmed et al. 2017). Due to this adhesion effect, the BUE from the tool tip adheres to the surface and consequently leaves such surface defects (BUE) (Liu et al. 2015). This defect is significantly reduced by the use of ultrasonic assistance, cf. Figure 10b, d and f. The surfaces machined with USAM of FeNi36 + 1% Zr exhibit a high number of cavities in contrast to the other two materials. These cavities are assumed to be pores that were formed during the welding process, cf. Sect. 3.1. Overall, USAM causes for all materials a significantly lower defect density of the finish milled surfaces compared to the conventional milling process.

4 Conclusion

The FeNi36 alloy was modified with 1 wt.-% of titanium or zirconium, respectively. The following influences on the welded microstructure and on the machinability were observed.

(1) PTA welding causes numerous finely distributed precipitates with high silicon content, occuring in the microstructure of FeNi36 as well as of both modified alloys.
(2) The modification with Ti and Zr both causes enhanced grain coarsening.
(3) In comparison to conventional finishing milling, USAM leads to a reduction of the arithmetic mean roughness *Ra* for all three materials.
(4) USAM shows beneficial effects regarding a significant reduction of machining induced surface defects compared to the conventional process especially in case of the alloy modifications.

The aim of these studies is to gain comprehensive knowledge in order to derive processing recommendations that will make the production and repair of

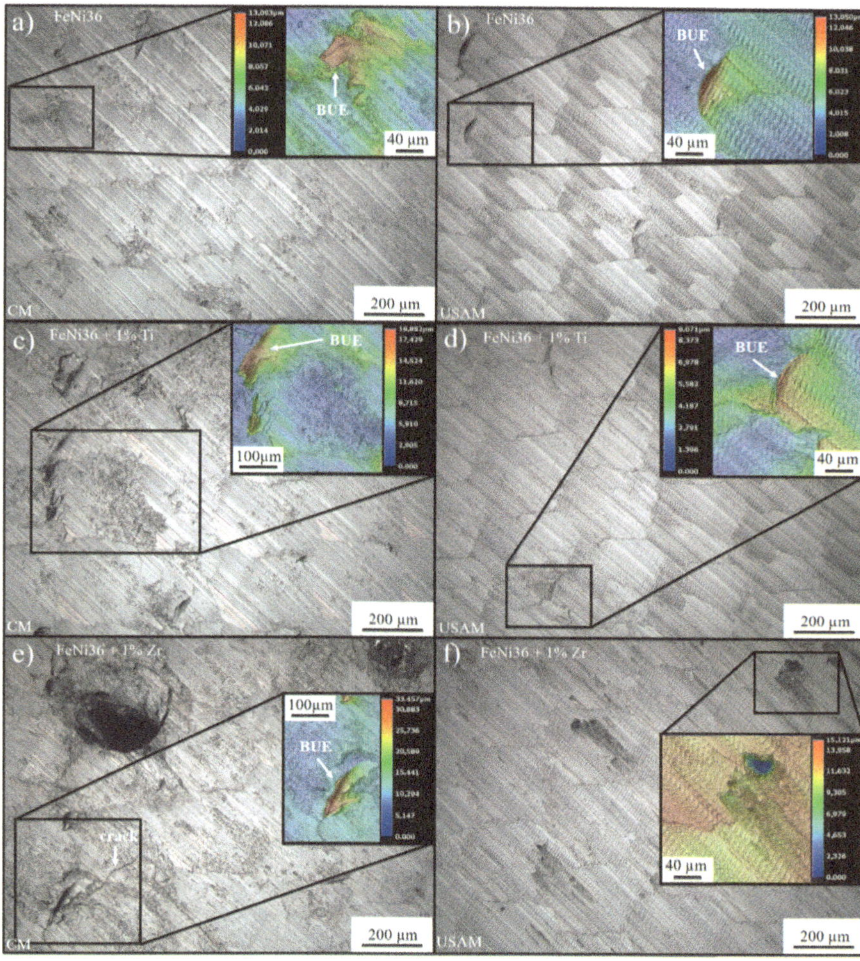

Fig. 10 Comparison of the topography of the FeNi36 initial alloy (**a**, **b**), FeNi36 + 1% Ti (**c**, **d**), FeNi36 + 1% Zr (**e**, **f**) and of conventional milling (**a**, **c**, **e**) and USAM (**b**, **d**, **f**)

these difficult-to-process alloys, especially for large components, significantly more efficient and economical.

Acknowledgements The IGF project IGF No. 20.979 N (DVS 01.3211) of the Research Association of the DVS was supported by the Federal Ministry for Economic Affairs and Energy by the AiF as part of the program for support of the cooperative industrial research (IGF) on the basis of a decision by the German Bundestag. We would like to thank the companies involved in the project committee for this funding and their support, in particular Corodur Fülldraht GmbH for providing the welding powder and WOLF Werkzeugtechnologie GmbH for providing the milling tools.

References

Abbasi, S.M., Morakabati, M., Mahdavi, R., Momeni, A.: Effect of microalloying additions on the hot ductility of cast FeNi36. J. Alloy. Compd. **639**, 602–610 (2015). https://doi.org/10.1016/j.jallcom.2015.03.167

Ahmed, Y.S., Fox-Rabinovich, G., Paiva, J.M., Wagg, T., Veldhuis, S.C.: Effect of built-up edge formation during stable state of wear in AISI 304 stainless steel on machining performance and surface integrity of the machined part. Materials (Basel) **10**(11) (2017). https://doi.org/10.3390/ma10111230

Ahmed, F., Ko, T.J., Kurniawan, R., Kwack, Y.: Machinability analysis of difficult-to-cut material during ultrasonic vibration-assisted ball end milling. Mater. Manuf. Processes 1–12 (2021). https://doi.org/10.1080/10426914.2021.1944194

Berns, H., Theisen, W.: Eisenwerkstoffe - Stahl und Gusseisen. Springer, Berlin, Heidelberg (2008)

Chen, C., Ma, B., Liu, B., He, J., Xue, H., Zuo, Y., Li, X.: Refinement mechanism and physical properties of arc melted invar alloy with different modifiers. Mater. Chem. Phys. **227**, 138–147 (2019). https://doi.org/10.1016/j.matchemphys.2019.02.006

DIN EN ISO 4287: Geometrische Produktspezifikation (GPS) – Oberflächenbeschaffenheit: Tastschnittverfahren – Benennungen, Definitionen und Kenngrößen der Oberflächenbeschaffenheit. Beuth publishing DIN (2010). https://doi.org/10.31030/1699310

Eissel, A., Engelking, L., Treutler, K., Schroepfer, D., Wesling, V., Kannengiesser, T.: Modification of Co-Cr alloys to optimize additively welded microstructures and subse-quent surface finishing. IIW-Doc No IX-NF-26-2021 (2021 under review)

Engelking, L., Schroepfer, D., Kannengiesser, T., Eissel, A., Treutler, K., Wesling, V.: Untersuchungsmethoden beim Fräsen additiv gefertigter schwer spanbarer Kobalt-Chrom-Legierungen. In: Tagung der Werkstoffprüfung 2021, Aachen, 2–3 December 2021 (2021 in press)

Guévenoux, C., Hallais, S., Charles, A., Charkaluk, E., Constantinescu, A.: Influence of interlayer dwell time on the microstructure of Inconel 718 Laser Cladded components. Opt. Laser Technol. **128** (2020). https://doi.org/10.1016/j.optlastec.2020.106218

Lagarec, K., Rancourt, D.G., Bose, S.K., Sanyal, B., Dunlap, R.A.: Observation of a composition-controlled high-moment/low-moment transition in the face centered cubic Fe–Ni system: Invar effect is an expansion, not a contraction. J. Magn. Magn. Mater. **236**(1–2), 107–130 (2001). https://doi.org/10.1016/s0304-8853(01)00449-8

Liu, C., Ren, C., Wang, G., Yang, Y., Zhang, L.: Study on surface defects in milling Inconel 718 super alloy. J. Mech. Sci. Technol. **29**(4), 1723–1730 (2015). https://doi.org/10.1007/s12206-015-0345-1

Nespor, D., Denkena, B., Grove, T., Pape, O.: Surface topography after re-contouring of welded Ti-6Al-4V parts by means of 5-axis ball nose end milling. Int. J. Adv. Manuf. Technol. **85**(5–8), 1585–1602 (2015). https://doi.org/10.1007/s00170-015-7885-5

Ni, C., Zhu, L., Liu, C., Yang, Z.: Analytical modeling of tool-workpiece contact rate and experimental study in ultrasonic vibration-assisted milling of Ti–6Al–4V. Int. J. Mech. Sci. **142–143**, 97–111 (2018). https://doi.org/10.1016/j.ijmecsci.2018.04.037

Rinck, P.M., Gueray, A., Kleinwort, R., Zaeh, M.F.: Experimental investigations on longitudinal-torsional vibration-assisted milling of Ti-6Al-4V. Int. J. Adv. Manuf. Technol. **108**(11–12), 3607–3618 (2020). https://doi.org/10.1007/s00170-020-05392-w

Sahoo, A., Medicherla, V.R.R.: Fe-Ni Invar alloys: a review. Mater. Today Proc. **43**, 2242–2244 (2021). https://doi.org/10.1016/j.matpr.2020.12.527

Salgueiriño-Maceira, V., Correa-Duarte, M.A., Duman, E., Farle, M.: FePt nanocrystals embedded in methylmethacrylate polymers. J. Magn. Magn. Mater. **299**(2), 467–471 (2006). https://doi.org/10.1016/j.jmmm.2005.05.007

Schroepfer, D., Treutler, K., Boerner, A., Gustus, R., Kannengiesser, T., Wesling, V., Maus-Friedrichs, W.: Surface finishing of hard-to-machine cladding alloys for highly stressed components. Int. J. Adv. Manuf. Technol. **114**(5–6), 1427–1442 (2021). https://doi.org/10.1007/s00170-021-06815-y

Sui, Q.S., Li, J.X., Zhai, Y.Z., Sun, Z.H., Wu, Y.F., Zhao, H.T., Feng, J.H., Sun, M.C., Yang, C.L., Chen, B.A., Peng, H.F.: Effect of alloying with V and Ti on microstructures and properties in Fe–Ni–Mo–C invar alloys. Materialia **8** (2019). https://doi.org/10.1016/j.mtla.2019.100474

Ulutan, D., Ozel, T.: Machining induced surface integrity in titanium and nickel alloys: a review. Int. J. Mach. Tools Manuf. **51**(3), 250–280 (2011). https://doi.org/10.1016/j.ijmachtools.2010.11.003

Weichmagnetische Eisen-Nickel-Legierungen. Auerhammer Metallwerk - Wickeder Gruop (2021). https://www.auerhammer.com/fileadmin/Data/Divisions/AMW/Download/deutsch/Werkstoffdatenbl%C3%A4tter/Weichmagnetische_Eisen-Nickel-Legierungen.pdf. Accessed 15 Oct 2021

Yu, Y., Chen, W., Zheng, H.: Effects of Ti-Ce refiners on solidification structure and hot ductility of Fe-36Ni invar alloy. J. Rare Earths **31**(9), 927–932 (2013). https://doi.org/10.1016/s1002-0721(12)60381-0

Zheng, X.W., Ying, G.F., Chen, Y., Fu, Y.C.: The Effects of cutting parameters on work-hardening of milling Invar 36. Adv. Mater. Res. **1089**, 373–376 (2015). https://doi.org/10.4028/www.scientific.net/AMR.1089.373

Maximisation of the Achievable Bond Width in Capacitor-Discharge-Welding of Annular Axial Seams on Gear Components

Julius Lindenmaier, Tim Hertzschuch, and Uwe Füssel

Abstract Capacitor-discharge-welding (CD-welding) is a highly dynamic process in which the entire weld seam cross-section is produced simultaneously by a high-current pulse of up to several hundred kiloampere. Welding tasks in gear manufacturing include diameters up to 250 mm and weld seam cross sections up to 1500 mm^2, which can be CD-welded in less than 40 ms. The fast welding process runs event-driven, it is not possible to intervene in the process sequence or to adjust parameters during welding. This means that the welding result is already defined by the boundary conditions and input parameters before welding. In this paper, the influence of various input parameters (current shape, electrode force, repositioning behavior and joining geometry) on the maximum achievable bond width in CD-welding of annular axial seams is investigated experimentally on a case hardening steel. The process limit is defined by the formation of macroscopic weld spatters, which can endanger the quality of the weld result. It is shown, that all the input parameters investigated have a significant influence on the achievable bond width. In particular, very short current rise times and low electrode forces severely limit the achievable bond width.

Keywords CD-welding · Joining gear components · Case hardening steel · Bond width · Quasi-static strength

J. Lindenmaier (✉)
Mercedes-Benz AG, Mercedesstraße 142, 70327 Stuttgart, Germany
e-mail: Julius_Enno.Lindenmaier@daimler.com

T. Hertzschuch · U. Füssel
Chair of Joining Technology and Assembly, Dresden University of Technology, George-Bähr-Straße 3c, 01069 Dresden, Germany
e-mail: tim.hertzschuch@mailbox.tu-dresden.de

U. Füssel
e-mail: uwe.fuessel@tu-dresden.de

L. F. M. da Silva et al. (eds.), *2nd International Conference on Advanced Joining Processes (AJP 2021)*, Proceedings in Engineering Mechanics,
https://doi.org/10.1007/978-3-030-95463-5_7

1 Introduction

CD-welding can be assigned to the group of resistance welding processes and is nowadays mainly used as a projection welding process (DVS 2016). Rotationally symmetrical gear components can be welded together by at least one of the joining partners having an annular projection contour, to achieve the current density concentration needed. The locally generated heat flux density $\dot{q}$ (Joule resistance heating) in the joining plane depends on the square of the welding current density j multiplied by the electrical resistance R_F of the joint (Fritz and Schulze 2015):

$$\dot{q} = \frac{\dot{Q}}{A_L} = j^2 \cdot R_F \tag{1}$$

Due to the local constriction of the conductive cross-section A_L by the projection, the current density j and the electrical resistance R_F become very high at the joint compared to the rest of the component. This is the reason, why the heat generation is strongly localized to the weld seam. The high currents of up to 1000 kA, required for the joining of the large cross sections, can be achieved by temporarily storing the required energy in a capacitor bank and then abruptly discharging it via a high-current pulse transformer. The resulting pulse shaped welding current heats the joint to the required temperature within a few milliseconds, before the heat can spread to the rest of the component (DVS 2016).

Compared with the currently most common welding process in gear manufacturing, the fusion welding with laser beam, the CD-welding process offers economic advantages (lower investment costs, simpler process diagnostics and control, elimination of process gases and filler metal, lower energy demand and reduced surface requirements). Furthermore, it also offers decisive process advantages. Due to very short welding times (<40 ms) and a simultaneous heating and cooling of the entire circumference, a low residual stress state is favored and no relevant thermal distortion is expected. By discharging a second current pulse through the joint, a tempering effect in the microstructure of the resulting weld can be achieved, without additional equipment needed (Petersmann 2004; DVS Forschungsvereinigung 2003).

Due to these process characteristics, CD-welding is predestined for joining tasks in the powertrain. Nevertheless, one searches the literature almost in vain for publications on the CD-welding of rotational components. Consequently, the existing potential is currently insufficiently exploited.

2 CD-Welding of Annular Axial Seams

When welding annular seams on rotationally symmetrical force-transmitting components, there are different possibilities for arranging the weld seam with respect to the axis of rotation of the joining partners. Depending on the available installation space, accessibility, position of the functional surfaces or load case, a certain arrangement of the weld seam to the axis of rotation may prove to be particularly favorable. Figure 1 shows the three possible main weld arrangements relative to the axis of rotation of the joining partners. They can be divided into the categories radial, axial and oblique.

In this paper, the welding of axial seams is investigated. An oversize x on one of the components is used as projection geometry. Figure 2 describes the CD-welding process of annular axial seams in more detail.

Contact situation before welding:

- First, the joining partner are positioned between the electrodes. The contact angle φ causes the components to align with each other as soon as the electrode force

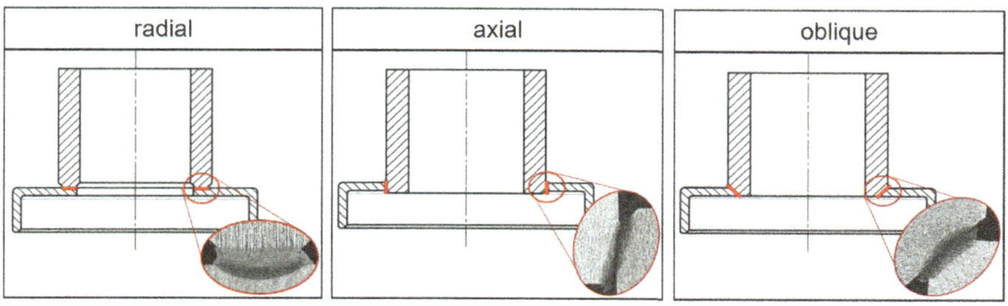

Fig. 1 Sectional view of different weld arrangements to the axis of rotation of the joining partners, using the example of a hollow shaft-hub connection

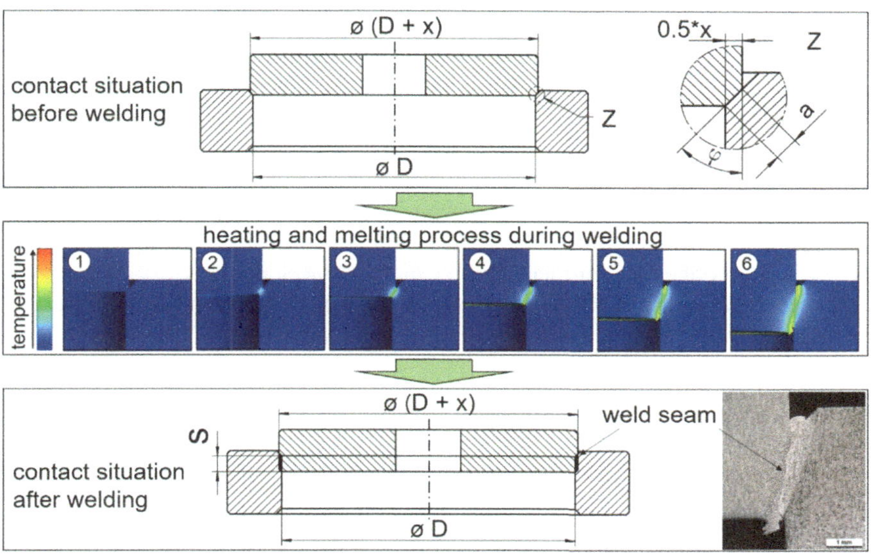

Fig. 2 Arrangement of the joining partners for CD-welding of annular axial seams

F is applied (self-centering). The resulting high pressure p on the joint already leads to an elastic and plastic deformation, resulting in the initial contact width a.

Heating and melting process during welding:

- The pulse-shaped welding current I leads to rapid temperature rise in the pressurized joint. As a result, the joining partners sink into each other (compare images 1–6 in Fig. 2) by the melt distance s and, after the end of the current flow, a materially bonded joint is formed.

Contact situation after welding:

- The resulting weld seam is never exactly axial, but has a slight inclination by a few degrees to the axis of rotation of the components, depending on the oversize x used. The achieved bond width is strongly dependent on the achieved melt distance s, but does not exactly equal it, because of the slight inclination and the resulting geometry on the corners of the seam.

This process variant was first described by Kniat (2008). However, fundamental investigations are very little researched for the production of this seam arrangement.

3 Problem Statement and Objective of Investigation

Weld seams on transmission components must be able to transmit dynamic moments and forces. In order to ensure the highest possible load-bearing capacity, the largest possible bond widths are required between the joining partners. Since the fast CD-welding process is event-driven, it is not possible to intervene or to adjust process parameters during welding. This means that the welding result (achieved melt distance s and bond width) is already defined by the boundary conditions and input parameters at the start of the process. Under otherwise constant boundary conditions and input parameters, only a certain amount of welding energy can be supplied into the process and a certain melt distance s (and therefore bond width) can be achieved before a macroscopic weld spatter formation occurs. This spatter formation represents a process limit, since it is hazardous to quality because it can damage finish-machined functional surfaces (e.g. tooth flanks, bearing seats) due to process heat or adhesion. Furthermore, it can lead to weld defects.

In recent years, different causes have been identified that lead to the formation of macroscopic weld spatters and defects in the weld seam during CD-welding:

- Loss of pressure or electrical contact (Stocks et al. 2017).
- Overheating and abrupt volume expansion (Ketzel et al. 2019; Lindenmaier et al. 2021).
- Breaking of the surface tension at the molten welding bulge (Lindenmaier et al. 2021).

The following main input parameters (except the charging energy E) that can be adjusted on a standard CD-welding machine and that affect these causes of macroscopic weld spatter formation are:

- The oversize x of the joining geometry.
- The rise time t_p of the welding current.
- The electrode force F.
- The spring stiffness k of the repositioning unit.

Objective of this investigation is therefore to examine the influence of these individual initial parameters on the maximum achievable bond width before a macroscopic weld spatter formation occurs.

4 Empirical-Experimental Approach

To study the influence of the different initial parameters on the maximum achievable bond width, only one parameter is varied at a time and the remaining parameters are kept constant. Several welding tests are then performed for each specific setting, in which the charging energy E of the capacitor is successively increased and the resulting melt distance s is measured in each case. The charging energy E at which the macroscopic weld spatter formation starts marks the process limit and the maximum achievable melt distance s. In order to verify sufficient strength of the welded joints, the welded samples are tested with a quasi-static press-out test (10 mm/min) and the maximum press-out force F_{po} is measured. Figure 3 shows the portal welding machine used and the setup of the press-out test.

The experiments are carried out on a welding diameter of 54 mm. The component geometry is described more detailed in Sect. 5.1. The component material used is the case-hardening steel 16MnCr5 + N (EN 10084 2008).

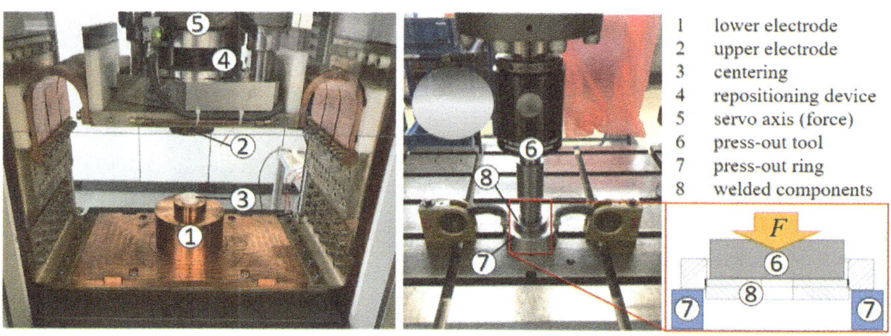

Fig. 3 Portal welding machine used (*left*) and setup of the quasi-static press-out test (*right*)

5 Experimental Work and Results

The results of the experimental work are presented below. Each input parameter is investigated within a separate chapter. First, the general conditions of the varied input parameter are described, before the welding results are presented. To visualize data points from tryouts with weld spatters occurring, the following determinations are made:

- Data points without filling: No weld spatter formation occurring.
- Data points with black filling: Slight weld spatter formation occurring.
- Data points with red filling: Strong weld spatter formation occurring (spatter limit).

5.1 Influence of the Oversize X

The oversize x determines the overlap between the joining partners. The larger the oversize x, the more material volume has to be melted to achieve a certain melting distance s. Furthermore, the oversize x influences the resulting contact width a before welding (comp. Figure 2) and thus the ohmic resistance R_F at the start of the process.

First, the actual size of the different oversizes x_1–x_3 (shaft) and of the hub diameter D_{hub} was measured for 10 samples of each geometry. The results and the general geometry of the joining components are shown in Fig. 4.

The geometric features show very slight deviations from the nominal dimensions and are within the specified tolerance. Next, the achievable melt distance s for the different oversizes x is tested and the influence on the press-out force F_{po} is investigated. The other input parameters stayed constant. The electrode force F was 60 kN, the current rise time t_p was 11.3 ms and the repositioning stiffness k was 8.5 kN/mm (spiral spring package). The results are shown in Fig. 5.

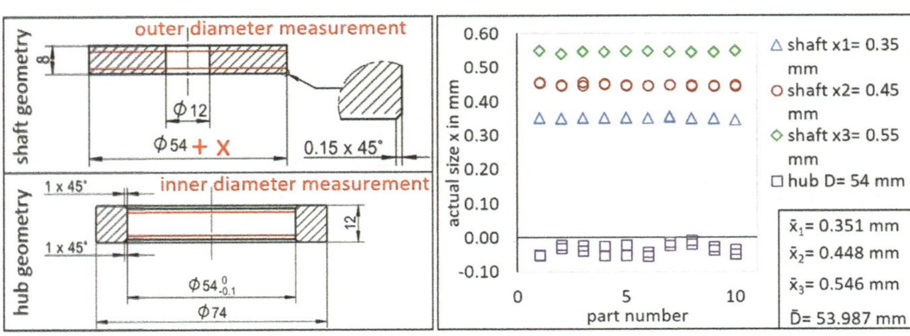

Fig. 4 General geometry of the joining components (*left*) and results of the measurement of the geometric features (*right*)

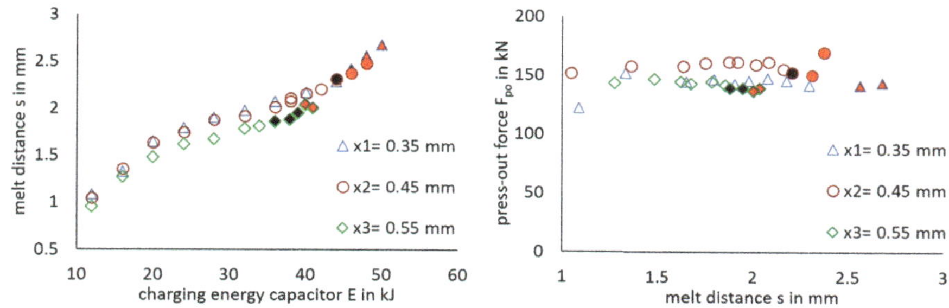

Fig. 5 Achievable melt distance for different oversizes (*left*) and results of the press-out test (*right*)

It can be recognized that too large oversizes x (here 0.55 mm) limit the achievable melt distance s (and therefore the bond width) before the weld spatter limit is reached. The weld samples for the oversize $x_2 = 0.45$ mm showed a slightly increased strength compared to oversizes x_1 and x_3 (comp. Figure 5, right).

5.2 *Influence of the Current Rise Time t_p*

The shape of the welding current influences the heat generation (comp. Equation 1) over time in the welding area (Ketzel et al. 2019). In addition, the current shape has a direct influence on the magnetic force effects on the molten welding bulge (Lindenmaier et al. 2021). The resulting welding current curve from a capacitor-discharge is essentially determined by the basic variables of the discharge circuit and cannot be actively controlled (Rusch 2018). For a higher charging energy E the peak current I_p rises, but the current rise time t_p and the welding time t_h remain almost constant. The current times can be adapted to the welding task, especially by adjusting the capacitance C of the capacitor-battery and the ratio $ü$ of the transformer.

The different machine-settings used and the resulting current rise times t_{p1}–t_{p4} are shown for a capacitor-discharge of 20 kJ in Fig. 6.

Small capacitances C and small transformer ratios $ü$ result in short current rise times t_p and high peak currents I_p.

Next, the achievable melt distance s for the different current rise times t_{p1}–t_{p4} is tested and the influence on the press-out force F_{po} is investigated. The other input parameters stayed constant. The electrode force F was 60 kN, the oversize x was 0.45 mm and the repositioning stiffness k was 8.5 kN/mm (spiral spring package). The results are shown in Fig. 7.

It can be recognized that too short current rise times t_p (here 5.5 ms) severely restrict the achievable melt distance s (and therefore the bond width) before the weld spatter limit is reached. In addition, a higher charging energy E of the capacitors is required for longer current rise times t_p to achieve the same melt distance s, as lower peak currents I_p result in a lower heat generation (comp. Equation 1). The current

CD of E= 20 kJ	t_{p1}	t_{p2}	t_{p3}	t_{p4}
capacitance **C** in mF	8	24	24	24
transformer ratio **ü**	1:120	1:120	1:160	1:240
rising time $\mathbf{t_p}$ in ms	5.3	8.8	11.3	15.6
welding time $\mathbf{t_h}$ in ms	11.5	16.8	21.6	30.1
peak current $\mathbf{I_p}$ in kA	351	281	232	183

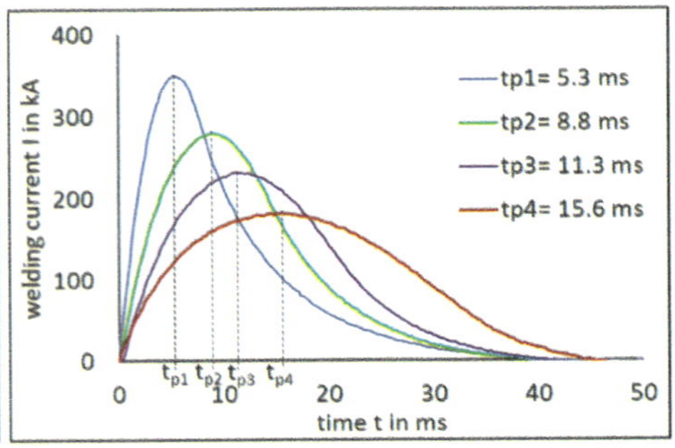

Fig. 6 Machine settings t_{p1}–t_{p4} (*left*) and resulting current times (*right*) for a capacitor-discharge of 20 kJ

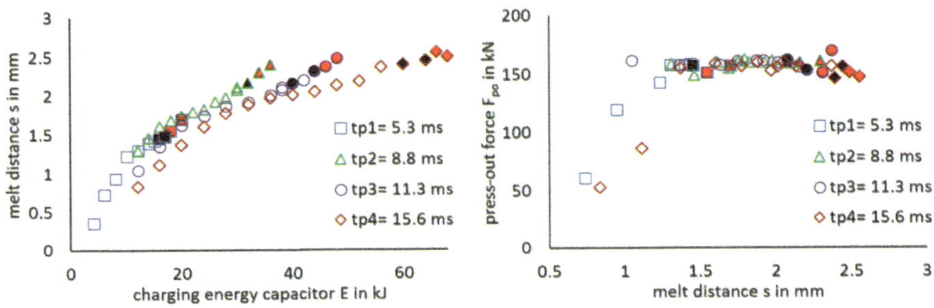

Fig. 7 Achievable melt distance for different current rise times (*left*) and results of the press-out test (*right*)

rise time t_p had no visible effect on the press-out force F_{po} needed to destroy the weld samples (comp. Figure 7, right).

5.3 *Influence of the Electrode Force* F

The electrode force influences the resulting pressure in the joining area during the welding process. In addition, the resulting deformations influence the ohmic resistance (especially the contact width a at the beginning of the welding process).

The contact widths a resulting from the plastic deformation and the electrode force curves over the welding time are shown for the electrode forces F_1–F_3 in Fig. 8.

Higher electrode forces F result in higher plastic deformations. The estimated pressure at the contact area is in the range between 700–800 MPa. The force drop ΔF during welding (comp. Figure 8, right) is a result of the relaxation of the spring package due to the melt distance s and depends on the spring stiffness k used.

Next, the achievable melt distance s for the different electrode forces F_1–F_3 is tested and the influence on the press-out force F_{po} is investigated. The other input

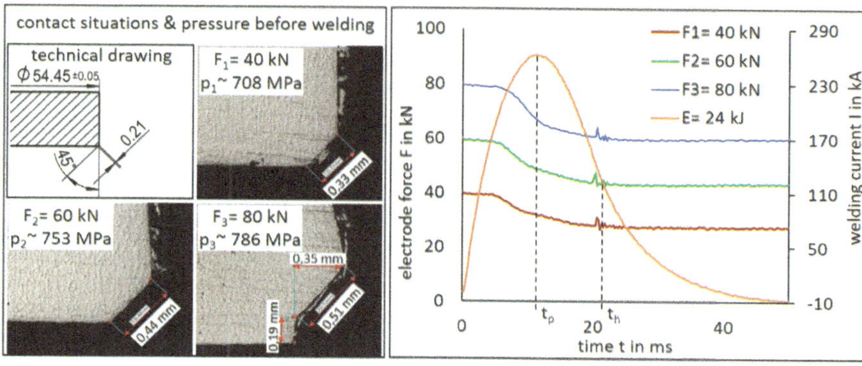

Fig. 8 Resulting deformation and estimated pressure at the contact before welding (*left*) and electrode force curves (*right*) for different electrode forces and a capacitor-discharge of 24 kJ

parameters stayed constant. The oversize x was 0.45 mm, the current rise time t_p was 11.3 ms and the repositioning stiffness k was 8.5 kN/ mm (spiral spring package). The results are shown in Fig. 9.

It can be recognized that too low electrode forces F (here 40 kN) severely restrict the achievable melt distance s (and therefore the bond width) before the weld spatter limit is reached. In addition, a higher charging energy E of the capacitors is required for lower electrode forces F to achieve the same melt distance s. For higher electrode forces F higher press-out force F_{po} are needed to destroy the weld samples (comp. Figure 9, right).

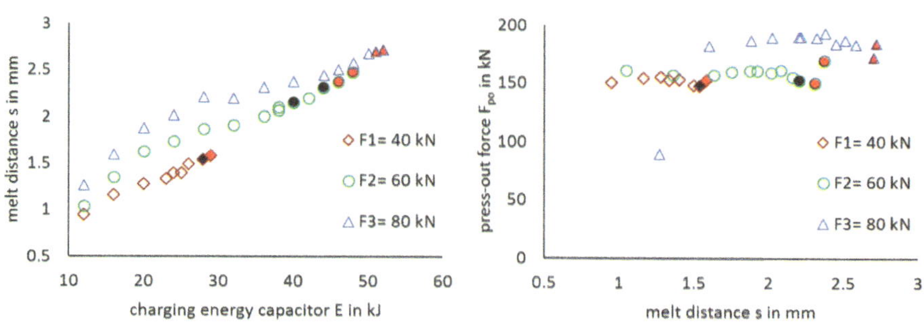

Fig. 9 Achievable melt distance for different electrode forces (*left*) and results of the press-out test (*right*)

5.4 *Influence of the Spring Stiffness* k *of the Repositioning Device*

In order to be able to continue transmitting the welding force and to avoid contact loss in the joining area or between the electrode and the component, at least one of the two electrodes must be able to follow the rapid softening projection geometry with as little delay as possible. Since the force generation system is often too sluggish, a repositioning device is positioned in the force flow which is decoupled from the force generation system (Stocks et al. 2017). The repositioning device is often designed in form of a spring package. The greater the spring stiffness k, the greater the loss of electrode force ΔF during welding for the same melt distance s.

There were four different parallel spring assemblies used which consist of several single springs. Two packages with spiral springs (linear characteristic) and two packages with elastomer springs (progressive characteristic). The spring stiffnesses k for the elastomer spring packages are estimated in the range of the electrode force around 60 kN:

- spiral spring package 1 with $k_1 = 12 \cdot 0.708$ kN/mm = 8.5 kN/mm.
- spiral spring package 2 with $k_2 = 12 \cdot 1.56$ kN/mm = 18.7 kN/mm.
- elastomer spring package 1with $k_3 = 6 \cdot 0.5$ kN/mm = 3 kN/mm.
- elastomer spring package 2 with $k_4 = 6 \cdot 1.67$ kN/mm = 10 kN/mm.

The spring characteristic curves and the resulting electrode speed during welding are shown for the different spring packages in Fig. 10.

The slowest spring package (elastomer spring 2) reaches a electrode speed v of 0.25 m/ s compared with 0.35 m/s for the fastest spring package (elastomer spring 1).

Next, the achievable melt distance s for the different spring packages is tested and the influence on the press-out force F_{po} is investigated. The other input parameters stayed constant. The electrode force F was 60 kN, the oversize x was 0.45 mm and the current rise time t_p was 11.3 ms. The results are shown in Fig. 11.

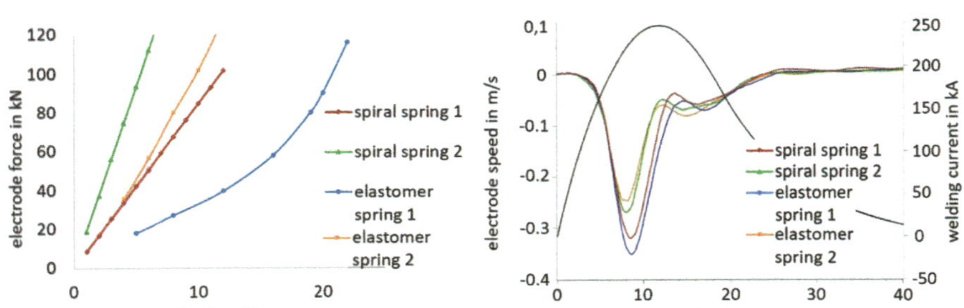

Fig. 10 Spring characteristic curves (*left*) and resulting electrode speed curves (*right*) for different spring packages and a capacitor-discharge of 32 kJ

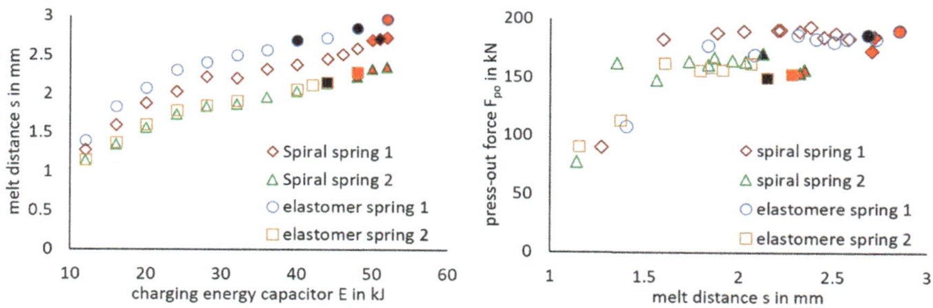

Fig. 11 Achievable melt distance for different spring packages (*left*) and results of the press-out test (*right*)

It can be recognized that a higher spring stiffness k and a lower resulting electrode speed v lower the achievable melt distance s (and therefore the bond width) before the weld spatter limit is reached. In addition, a higher charging energy E of the capacitors is required for a higher spring stiffness k to achieve the same melt distance s. For lower spring stiffnesses k and a higher resulting electrode speed v higher press-out forces F_{po} are needed to destroy the weld samples (comp. Figure 11, right).

6 Discussion of the Results

Too large oversizes x lead to a premature weld spatter formation because of the large amount of molten metal that is squeezed-out from the seam into the welding bulge. Too short weld times t_p lead to a premature weld spatter formation because of the high resulting peak currents I_p that lead to a high influence of the Lorentz force and breaking of the surface tension on the molten welding bulge (Lindenmaier et al. 2021). Low electrode forces F lead to a premature weld spatter formation because of a lack of pressure or a contact loss between the components. High spring stiffnesses k and a slow resulting electrode acceleration and speed lead to a higher force loss ΔF and therefore to an earlier weld spatter formation occurring (Stocks et al. 2017).

All weld samples with melt distances s greater than 1.5 mm resulted in press-out forces F_{po} over 140 kN and a fracture in the base material outside the heat affected zone. The higher strength for high electrode forces F (comp. Figure 9, right) and for low spring stiffnesses k (comp. Figure 11, right) must be due to a resulting work hardening in the base material around the weld seam, as consequence of the high pressure. Figure 12 shows samples of the fracture after the press-out test as well as a hardness curve through the heat-affected zone (HAZ).

It can be observed, that the samples typical crack outside the harder HAZ in the softer base material. This indicates a high seam quality. In addition, the width of the narrow HAZ was less than 1.5 mm.

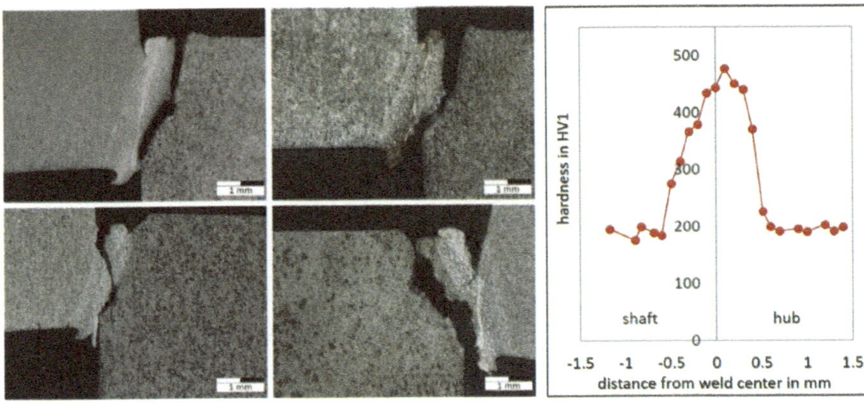

Fig. 12 Fracture in the base material after the press-out test (*left*) and hardness measurement through the heat affected zone of the shaft-hub weld (*right*)

7 Conclusion

Components with a welding diameter of 54 mm made from the case hardening steel 16MnCr5 (EN 10084 2008) were investigated. It was shown, that all the input parameters investigated had a significant influence on the achievable weld bond width. In particular, very short current rise times t_p and low electrode forces F severely limited the achievable bond width. For the given boundary conditions, an oversize x around 0.45 mm (see Fig. 5), a current rise time t_p around 11 ms (see Fig. 7), an electrode force F around 70 kN (see Fig. 9) and a spring stiffness k around 3 kN/ mm (see Fig. 11) proved to be the best input parameters to achieve a maximum bond width, before a macroscopic weld spatter formation occurs.

In general, melt distances s between 2 and 3 mm could be achieved, without a macroscopic weld spatter formation occurring. For a welding diameter of 54 mm, this results in a bonded cross section of 340–510 mm^2. For an average press-out force F_{po} of 175 kN and an average bonded cross section of 425 mm^2, this results in a calculated shear strength τ of 412 N/ mm^2 of the weld seam.

Acknowledgements The investigations presented were carried-out in cooperation with the Mercedes-Benz AG and were financially supported.

References

Deutscher Verband für Schweißen und Verwandte Verfahren e.V.: Merkblatt DVS 2911, Kondensatorentladungsschweißen - Grundlagen, Verfahren und Technik. DVS Media GmbH, Düsseldorf (2016)

EN 10084: Case hardening steels-technical delivery conditions (2008)

Fritz, A.H., Schulze, G.: Fügen. In: Fertigungstechnik, pp. 125–272. Springer Vieweg, Berlin, Heidelberg (2015)

Forschungsvereinigung, D.V.S.: Widerstandsschweißen von höherkohlenstoffhaltigen Stählen mit sehr kurzer Wärmeeinbringung. DVS Media GmbH, München (2003)

Ketzel, M.-M., Zschetzsche, J., Füssel, U.: Erwärmungsverhalten in der Kontaktzone beim Kondensatorentladungsschweißen unter Berücksichtigung der dynamischen Stromänderung und des Nachsetzverhaltens der Elektroden. Schweißen Und Schneiden **71**(1–2), 24–34 (2019)

Lindenmaier, J., Lindenau, D., Bachker, P., Füssel, U.: Einfluss der Magnetfeldverteilung auf die Schweißnahtausbildung beim Kondensatorentladungsschweißen von Getriebebauteilen. Paper presented at Große Schweißtechnische Tagung 2021: DVS Kongress 2021, Deutscher Verband für Schweißen und Verwandte Verfahren e.V., Essen (2021)

Petersmann, C.: Kondensator-Impulsschweißen höher kohlenstoffhaltiger, randschichtgehärteter Einsatzstähle am Beispiel eines Pkw-Sitzbeschlages. Dissertation, Technische Hochschule Aachen (2004)

Rusch, H.-J.: Prozesstechnische Anforderungen und elektrotechnisches Design von Kondensatorentladungsschweißmaschinen. Dissertation, Technical University of Dresden (2018)

Stocks, N., Ketzel, M.-M., Zschetzsche, J., Füssel, U.: Optimierungspotentiale beim Kondensatorentladungsschweißen durch maschinendynamische Auslegung der Nachsetzeinheit. Paper presented at Große Schweißtechnische Tagung 2017: DVS Kongress 2017, Deutscher Verband für Schweißen und Verwandte Verfahren e.V., Düsseldorf (2017)

Process Comparison of Friction Stir Welding, MIG, Laser Beam Welding and Laser Hybrid Welding in Joining Aluminium EN AW-6063 T6

Iurii Golubev and Axel Meyer

Abstract Four different welding processes had been investigated and compared to each other in joining aluminium extrusion for a battery tray panel in the automotive industry. Friction Stir Welding (FSW), Metal Inert Gas Welding (MIG), Laser Beam Welding (LBW) and Laser MIG Hybrid Welding (LMH). Weld properties had been analysed by metallographic cross sections, tensile tests and fatigue tests after initial visual inspection. Simultaneously to the mechanical properties data for a carbon footprint evaluation had been gathered as well. The energy consumption, feed rates for welding consumables and shielding gas flow which had been analysed were appropriate. The GaBi software database for life cycle assessment was used for final evaluation. Significant differences had been identified proving tremendous differences for a carbon neutral production.

1 Introduction

The high use of aluminum in the automotive industry, especially for EV battery trays, results in an increasing demand for welding of aluminum alloys. These welded joints often made of hollow extrusions have to meet high requirements in terms of their tightness, strength, and crash behavior. In addition, it should be possible to fully automate a welding process to ensure high reproducibility of the welded parts. Thus, the selection of welding methods for the production of the battery trays is a challenging task.

To weld aluminum extrusions for a battery tray panel, the automotive industry currently prefers friction stir welding (FSW) (Meyer 2019), which ensures excellent quality and high reproducibility of the welded joints and does not require expensive weld preparation. Despite the advantage of FSW, competing welding methods should also be considered. Since laser beam welding (LBW) has been widely adopted into the automotive industry due to its high processing speed and process flexibility, it

I. Golubev · A. Meyer (✉)
Mercatorstraße 65b, 21502 Geesthacht, Germany
e-mail: axel.meyer@riftec.de

L. F. M. da Silva et al. (eds.), *2nd International Conference on Advanced Joining Processes (AJP 2021)*, Proceedings in Engineering Mechanics,
https://doi.org/10.1007/978-3-030-95463-5_8

can nowadays also be used for welding of the battery tray panels (TRUMPF 2021). A further welding method now used for battery tray production is laser-MIG hybrid welding (LMH) (Fronius 2020), which offers the gap-bridging ability and easy weld-seam preparation of MIG welding, as well as the low heat input, deep penetration, and speed of laser welding. Although FSW, LBW and LMH have proven to give quality welded joints used in automobile fabrication, their application is limited by the high cost of equipment. Therefore, the arc welding method such as metal inert gas welding (MIG) with its low cost compared to other welding techniques and high energy efficiency, therefore, remains a useful welding process in the automotive (Ogbonna et al. 2019).

To choose the adequate welding process, the processes should be compared to each other regarding quality, cost and time as well as the environmental friendliness. The quality of the welded joints made by different welding methods can be evaluated by visual inspection, macroscopic examination, tensile and fatigue tests. In order to get reliable results, all the welding processes should be tested for same joints typology and equal material. Evaluation of the cost and time is based on RIFTEC's and HAI's many years of experience in welding of aluminum alloys as well as the earlier research. Environmental impact of the different welding processes has been evaluated using the life-cycle assessment methodology (LCA).

2 Materials and Methods

In this study, $1200 \times 400 \times 12$ mm hollow extrusions made of EN AW-6063 T6 were selected as a base material. The chemical composition and mechanical properties of the base material are given in Table 1 and Table 2 respectively.

Four single sided butt welds and one double sided butt weld have been produced by each welding method. The desired weld penetration was 3 mm. The specific welding parameters used for each welding method are described below.

FSW was performed at HAI Ranshofen by using a double-spindle FSW machine integrated in a fully automated line for serial manufacturing of battery tray panels. No special treatment or optimisation was applied to the welded parts. Hence these parts illustrate the serial conditions on a real industrial level. Two hollow extrusions were welded simultaneously from both sides at a welding speed of 3 m/min.

Table 1 Chemical composition of the base material

Si	Fe	Cu	Mn	Mg	Cr	Zn	Ti	Al
0.47	0.2	0.02	0.05	0.47	0.01	0.02	0.01	Rest

Table 2 Mechanical properties of the base material

$Rp_{0.2}$, N/mm^2	Rm, N/mm^2	A 50%
222.8	244.8	12.7

Unlike FSW, LMH, LBW and MIG trials have been performed at scientific institutions by using universal welding machines and fixtures. In addition, less time was available for the selection of the welding parameters and process optimization. Therefore, welding parameters and welding fixtures used for fusion welding methods in this work were not applicable for a series production.

LMH was performed at SLV Halle GmbH. The welding was carried out by using a high-power diode laser LASERLINE LDF 15,000-60 at a welding speed of 2.2 m/min, the MIG welding power source QINEO PULSE 600, Argon gas at flow rate of 18 l/min, and filler wire was AlSi5 1.2 mm diameter.

LBW was performed at the Chair of joining and welding technology Brandenburg University of Technology (BTU) Cottbus-Senftenberg. The welding was carried out by using the laser machining center Reis Robotics equipped with 15 kW fiber laser IPG Photonics at a welding speed of 1.5 m/min, Argon gas at flow rate of 20 l/min, and filler wire was AlMg4,5Mn 0.8 mm diameter.

MIG was performed at the Chair of joining and welding technology Brandenburg University of Technology (BTU) Cottbus-Senftenberg. The welding was carried out by using the power source Fronius TPS 500i at a welding speed of 0.9 m/min, Argon gas at flow rate of 15 l/min, and filler wire was AlMg4,5Mn 1.2 mm diameter.

Welding methods have been compared regarding the quality, cost, time, and environmental friendliness. Each welding method can score from one to five points for each criterion. Evaluation results are summarized using radar charts.

To evaluate the quality of the welded joints, visual inspection, macroscopic examinations, tensile test, and fatigue test were carried out. The specimens for destructive testing were cut from two welded joints made by each welding method. Tensile test and fatigue test specimens have been prepared according to the standard DIN EN 50,125:2009-07 (Form H). All fatigue test specimens (14 pcs) and first set of tensile test specimens (6 pcs) have been machined on both sides to avoid the influence of the weld surface roughness on the fatigue und tensile test results. Second set of the tensile test specimens (6 pcs) has been machined on the side of the weld root only. Bending test specimens (6 pcs) have been prepared according to the standard DIN EN ISO 7438. Three samples each were taken from the beginning, middle and the end of the weld seam for macroscopic examination. Figures 1 and 2 illustrate the position of the tensile test and fatigue test specimens.

Environmental impact of the different welding processes has been evaluated using the life-cycle assessment methodology (LCA). Within a LCA the quality of the underlying data on materials, processes, energy, and transport is crucial to produce valuable and reliable results. While most LCA database providers source their data from aging literature or laboratory research-based content, LCA database used in this research (GaBi software) is built on primary industry data and in close cooperation with associations, providing a reliable environmental data foundation.

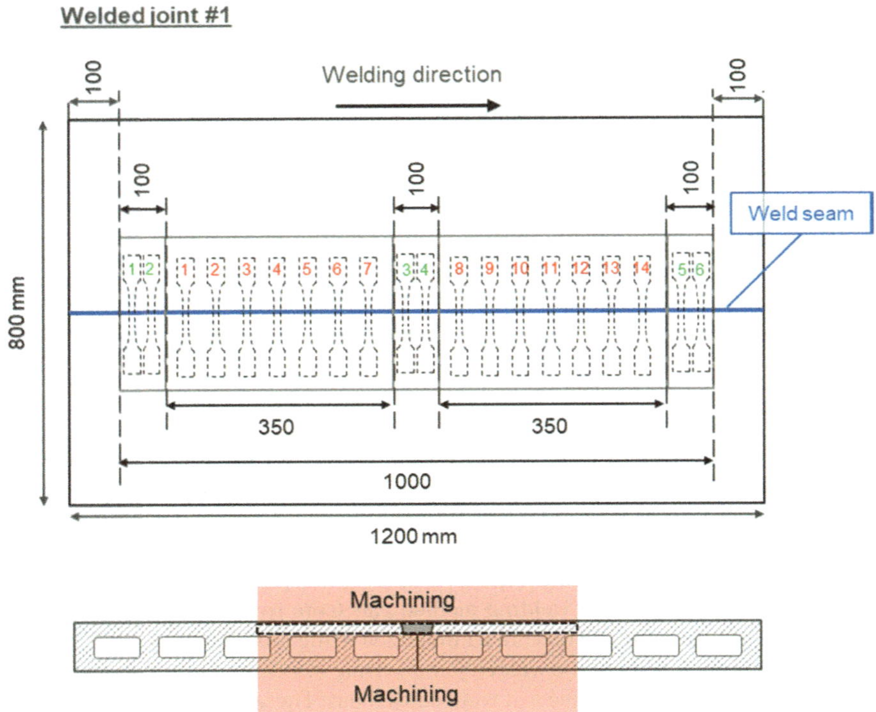

Fig. 1 Position of the tensile test and fatigue test specimens

3 Results and Discussions

Figure 3 shows the surface appearance of the welds made by FSW, LMH, LBW and WIG welding respectively.

FSW produces a relatively smooth surface of the weld seam without spatter and excess weld metal above the height of the parent metal. However, some of the material may be pushed out as toe flash on the workpiece surface at the edge of the weld seam. The width of the FSW weld remains constant along the welding line and corresponds to the tool shoulder diameter. In contrast to joint welded by FSW, joints welded by fusion welding methods exhibit a rougher weld surface, spatter on the workpiece surface, excess weld metal as well as irregular weld width.

Four types of surface welding imperfections were found on the welded parts. In order to measure "sensitivity" of the welding process to the occurrence of specific surface imperfections, the number of the welds affected by specific imperfection has been counted. Results have been summarized in the diagram (Fig. 4).

Since spatter and toe flash have a similar appearance i.e. excess material on the surface has to be removed by grinding or machining, they belong to one types of surface imperfections. Spatter have been found on all the joints made by LMH, LBW,

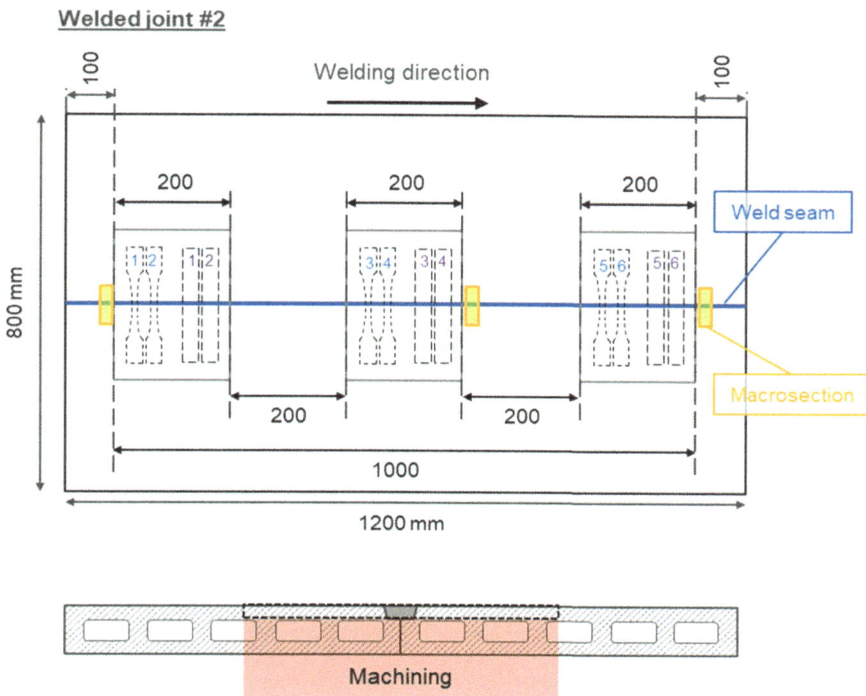

Fig. 2 Position of the tensile test specimens, bending test specimens, and metallographic cross-sections

and MIG, while toe flash was presented only on the one FSW joint. The second most frequently occurring surface imperfection is undercut found on LBW (four welds), LMH (three welds), and MIG (one weld) joints. Three MIG joints and one LMH joint exhibit surface porosity. Surface longitudinal cracks have been found on the welds made by LMH (one weld) and LBW (one weld). Thus, FSW has the lowest sensitivity to the occurrence of specific surface imperfections.

Figure 5 shows macroscopic sections taken from the middle of the weld seams.

The evaluation of the imperfections has been carried out according to ISO 25239-5:2020 (FSW) and DIN EN ISO 6520-1 (fusion welding methods). The FSW joint exhibits a high degree of continuity and no internal imperfections. In addition, it is apparent in this macrograph that FSW has good gap-bridging ability. LMH weld has also no internal imperfections. However, the weld depth was not constant along the weld line and greater than required. Joints welded by LBW exhibit cracks in the weld metal and porosity. MIG presents porosity as well as excess weld metal. Moreover, the weld depth was not constant and less than required.

Figure 6 shows the results of the tensile tests for (a) samples machined on both sides and (b) samples machined on one side.

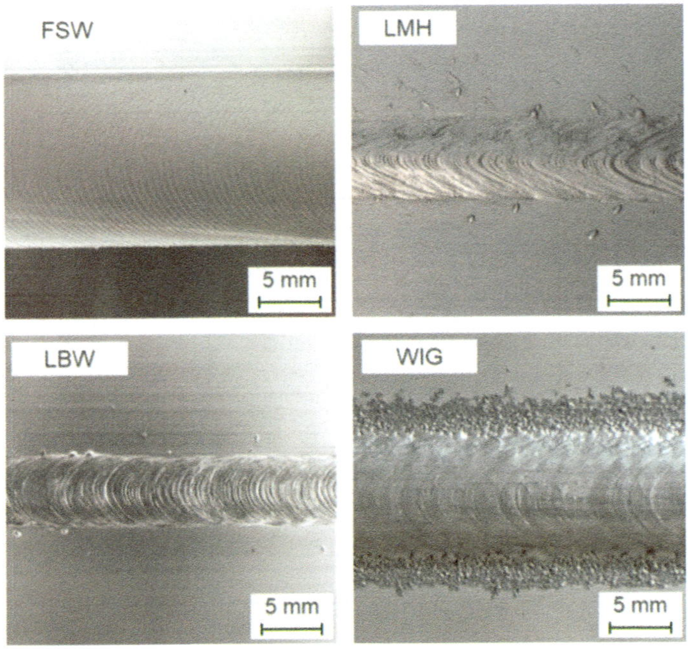

Fig. 3 Appearance of the joint welded by **a** FSW, **b** LMH, **c** LBW and **d** MIG

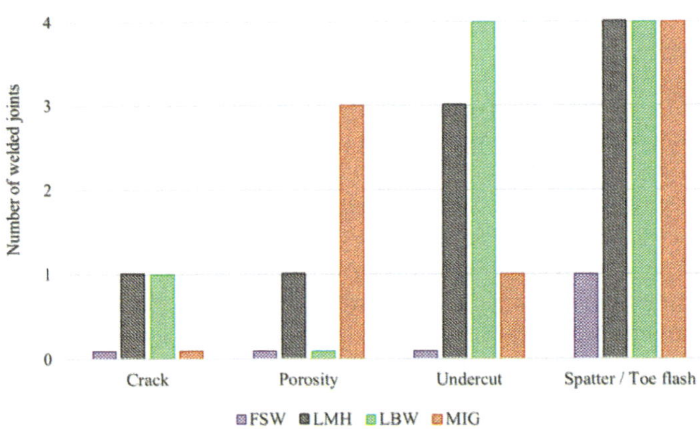

Fig. 4 Frequency of occurrence of welding defects

Results of the tensile tests show a similar behaviour for both sets of samples. As mentioned above, FSW produces welds without defects associated with fusion welding. Moreover, FSW creates fine recrystallized microstructure in the weld

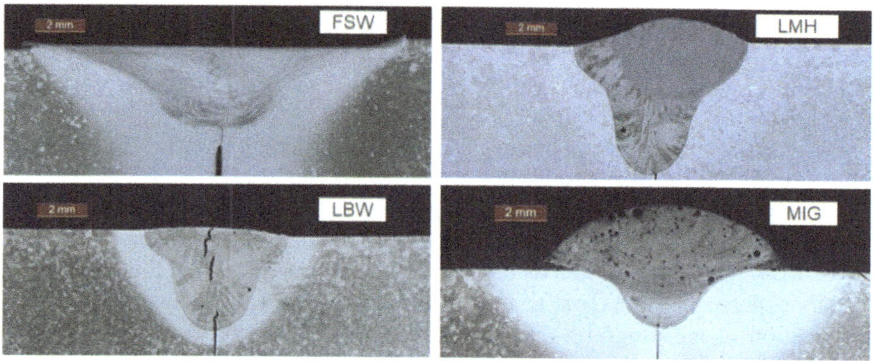

Fig. 5 Cross sections **a** FSW, **b** LMH, **c** LBW and **d** MIG

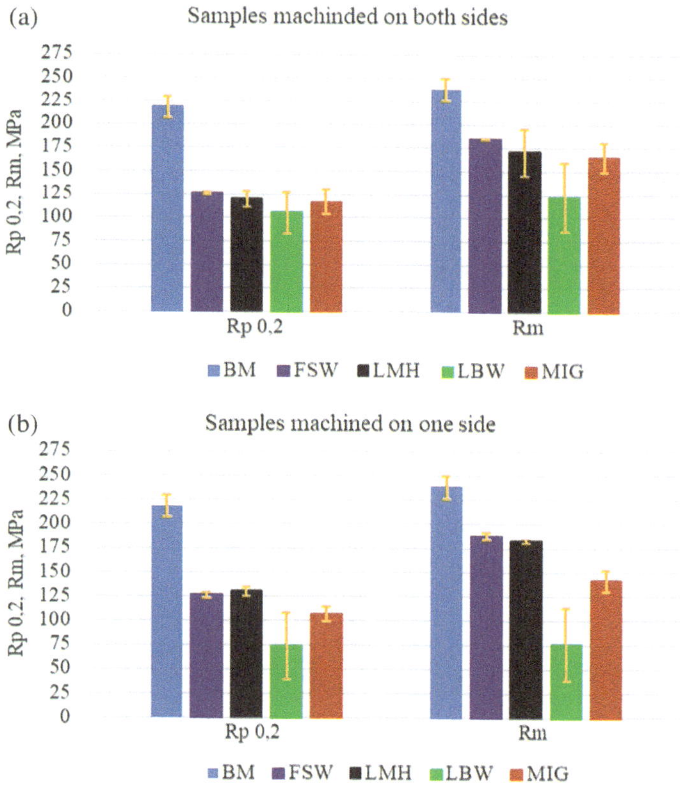

Fig. 6 Tensile test results for joints welded by FSW, LMH, LBW, MIG and BM. **a** Samples machined on both sides **b** Samples machined on one side

nugget. This results in excellent mechanical properties of its joints. Mechanical properties of the LMH joints are located at a slightly lower level as the FSW. LBW joints exhibits the lowest mechanical properties due to presence of the internal imperfection in its joints such as cracks and porosity.

Fatigue test was carried out in a resonant testing machine RUMUL MIKROTRON with a stress ratio R of 0.1 (pulsating tensile load) and a test frequency of 99 Hz. For each welding process, 14 fatigue test specimens (DIN EN 50,125:2009-07, Form H) were used to create S–N (stress-life) curve. Results of the fatigue test for FSW, LMH, LBW, and MIG, as well as base material (BM) are shown in Fig. 7.

Large spread in the fatigue test results for LMH and LBW joints do not allow the creation of meaningful S–N curves for these welding processes. Apparently, the reason for this is the presence of internal imperfection in these joints such as cracks and porosity. However, the stress amplitude can be evaluated to compare them with other welding processes. LBW and LMH joints exhibit the lowest fatigue strength. FSW joints show a considerably better fatigue performance than other welding methods and their S–N curve is similar to base material level.

In order to measure how environmentally friendly each welding process is, greenhouse gas emissions expressed as carbon dioxide equivalent (carbon footprint) have been calculated by using LCA with GaBi Software database. The results for the four welding processes are shown in Fig. 8 as CO_2 equivalent per meter weld length.

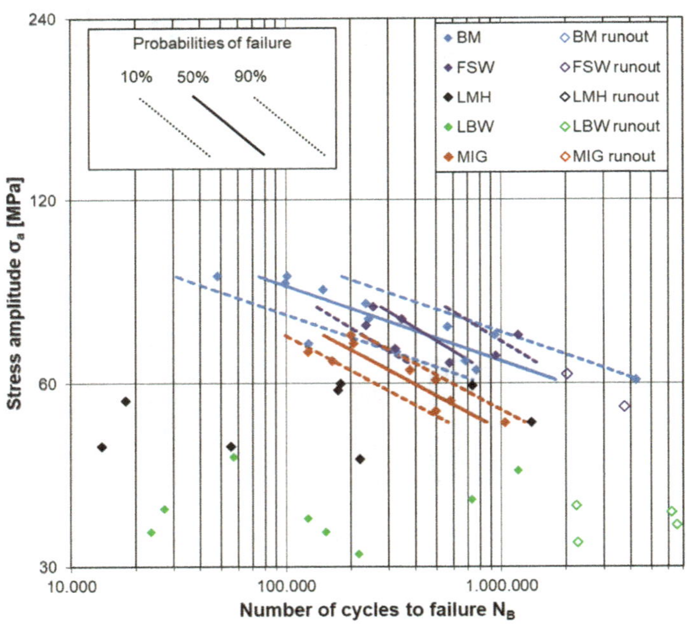

Fig. 7 S–N curves for joints welded by FSW, LMH, LBW, MIG and BM

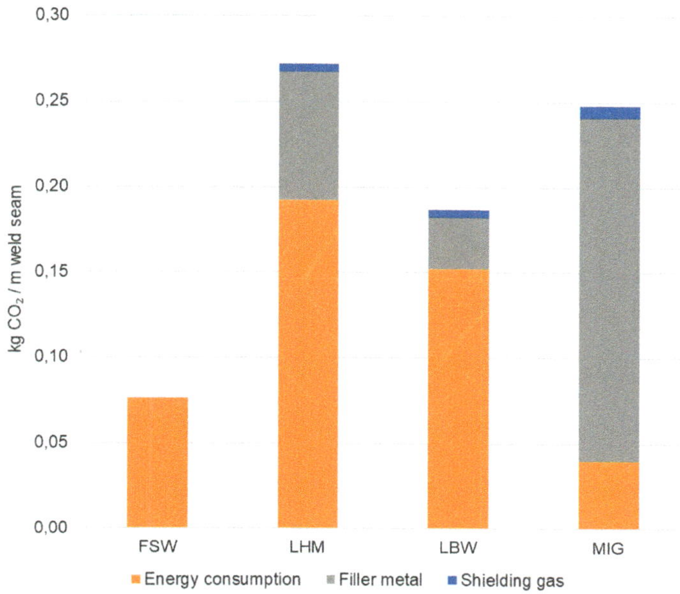

Fig. 8 Carbon footprints

The carbon dioxide (or greenhouse gas) emission in welding arises primarily from three sources: (1) energy consumption, (2) filler metal, and (3) shielding gas. If we consider the energy consumption impact only, MIG is the most environmental-friendly welding process. However, the high consumption of filler metal in MIG welding results in a fivefold increase in carbon dioxide emission. Laser-based welding processes require significantly less filler metal but consume more energy than FSW and MIG processes. The fact that FSW consumes relatively little energy and does not involve the use of filler metal and shielding gas provides it with an advantageous position over its competitors. In this analysis, we used an average Germany's electricity generation mix to evaluate CO_2 emissions from energy consumption. If CO_2-neutral energy was used, the energy consumption contribution to carbon footprint would be almost eliminated for all welding processes.

Based on the results described above, RIFTEC's and HAI's years of experience in welding of aluminum alloys as well as the earlier research, the comparison of the welding processes regarding quality, cost, time, and environmental friendliness are all summarized in radar charts (Fig. 9).

Quality. Since FSW is a solid-state joining technique, its welds can be produced with absence of solidification cracking, porosity, undercut, and other defects of fusion welding. This in return results in high mechanical and fatigue properties of the joints welded by FSW. In addition, a relatively low heat input during FSW leads to lower distortion.

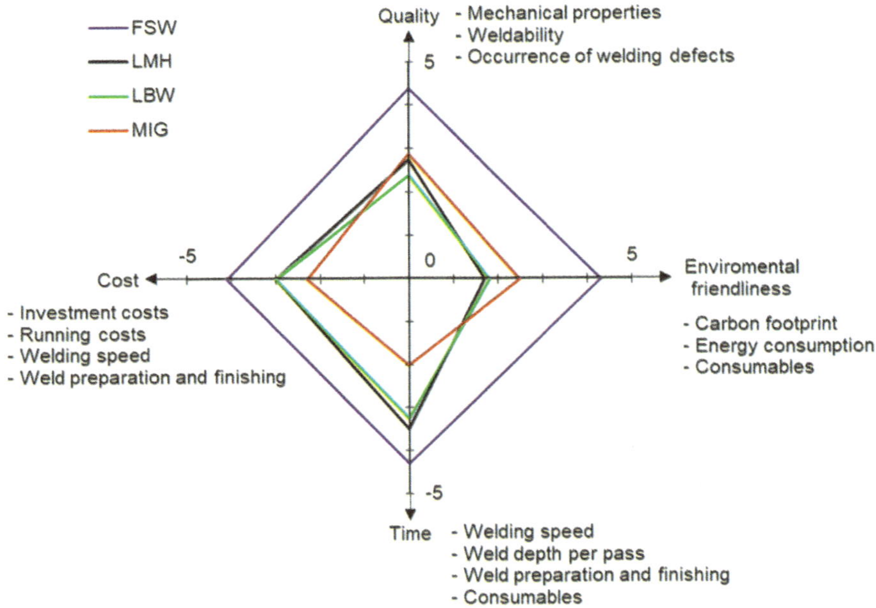

Fig. 9 Comparison of the welding processes

Environmental friendliness. As mentioned above, MIG consumes considerably less energy than other welding methods. However, shielding gas and great amounts of filler metal used during MIG produce a lot of greenhouse gases vented to the earth's atmosphere. Thus, MIG welding is not at all "green" technology. During FSW, no consumables such as a shielding gas or filler metal are used, and no harmful emissions such as radiation or metal fumes during fusion welding methods are created. All of this makes FSW more environmentally friendly than competing processes.

Time. The time has been evaluated as the total time spent on the weld preparation, welding, and weld finishing. MIG and LMH processes require the edge preparation prior to welding. In addition, MIG has the slowest welding speed and the lowest weld penetration depth in single pass, which make MIG the most time-consuming welding process. FSW is the fastest process regarding welding speed. Moreover, FSW produces a relatively smooth welding surface without spatter and excess weld metal. This results in a less time for surface finishing of the FSW welds.

Cost. The cost is presented as cost per part and includes investment cost (welding machine and fixtures), consumables (energy, shielding gas, filler metal), and manufacturing cost (productivity, weld preparation, finishing). LMH and LBW require high investment costs due to the expensive welding machines, control systems and safety devices. In addition, these processes consume more energy and require the use of shielding gas and filler metal. However, a higher productivity and less time-consuming weld preparation make LMH and LBW more cost effective than MIG (despite more lower investment cost for MIG). FSW has also high investment costs but lower energy consumption than LMH and LBW. Moreover, it does not require weld preparation and consumables such as a shielding gas or filler metal. Thus, FSW becomes the most cost-effective welding process.

4 Conclusions

In this work FSW, WIG, LBW and LMH have been investigated and compared to each other regarding their quality, cost, time, and environmental friendliness. All the welding processes were tested for same joints typology and equal material. FSW was performed by using a FSW machine integrated in a fully automated line for serial manufacturing. Unlike FSW, competing welding processes have been carried out at scientific institutions by using universal welding machines and fixtures. The quality of the welded joints made by different welding methods have been evaluated by visual inspection, macroscopic examination, tensile and fatigue tests. Evaluation of the cost, time, and environmental friendliness was based on the RIFTEC's and HAI's years of experience in welding of aluminum alloys as well as the earlier research. FSW showed the best result for all the comparison criterions. This work is a starting point and invitation for additional input from the welding community. A round robin have been started on basis of these initial data points.

References

Fronius International GmbH.: Laserhybrid welding excels when extrusion press profiles. http://www.fronius.com.tw/en/welding-technology/info-centre/press/ast-alu-menziken (2020). Accessed 22 November 2021

Meyer, A.: Series production of friction stir welded battery trays. In: Paper Presented at the 6th International Conference on Scientific and Technical Advances on Friction Stir Welding & Processing, Université catholique de Louvain, Louvain-la-Neuve (2019)

Ogbonna, O.S., Akinlabi, S.A., Madushele, N., Mashinini, P.M., Abioye, A.A.: Application of MIG and TIG welding in automobile industry. J. Phys. Conf. Ser. **1378**, 1–10 (2019)

TRUMPF GmbH+Co.KG.: Battery packs—laser welding and laser cleaning. https://www.trumpf.com/en_INT/solutions/industries/automotive/e-mobility/battery-packs-laser-welding-and-laser-cleaning. Accessed 22 November 2021

Joining High Nitrogen Steels with Ag–Cu–Ni Metal Filler

Shuye Zhang, Peng He, Zhenfeng Li, Xingxing Wang, Di Gao, and Dusan P. Sekulic

Abstract This paper offers a study of a vacuum brazing of high nitrogen steel by using AgCuNi metal filler. Moreover, the precipitation behavior of the second phase of the high nitrogen steel has been analyzed. The result shows that the temperature range of the second phase precipitates has been 700–800 °C. The heat treatment of high nitrogen steel base metal should be avoided in that temperature range. However, it has been established that the heat treatment of the brazing process has little influence on microstructures and mechanical properties of high nitrogen steel, thus, the brazing temperature range was established to be 850–1050 °C. A typical joint microstructure is mainly the Ag–Cu eutectic in the filler layer, and the element diffusion area near the high nitrogen steel side. Brazed joints, with good mechanical properties were obtained at temperatures of 850 °C and 950 °C. It has been established that the optimized brazing process conditions are: (i) the peak temperature of T = 950 °C and (ii) the dwell time at the peak of t = 20 min. The element diffusion processes of high nitrogen steel brazing have been considered by using molecular dynamics simulations. The diffusion processes of Fe–Cu and Fe–Ni binary systems were simulated. It has been uncovered that in the Fe–Cu diffusion process, only the atoms diffuse mutually. A mesophases were found during the Fe–Ni diffusion process.

Keywords High nitrogen steel · Precipitation in the second phase · Vacuum brazing · Microstructure · Molecular dynamics simulation

S. Zhang (✉) · P. He (✉) · Z. Li · D. Gao
State Key Laboratory of Advanced Welding and Joining, Harbin Institute of Technology, Harbin 150001, China
e-mail: syzhang@hit.edu.cn

P. He
e-mail: hepeng@hit.edu.cn

X. Wang
School of Mechanical Engineering, North China University of Water Resources and Electric Power, Zhengzhou 450045, China

D. P. Sekulic
College of Engineering, University of Kentucky, Lexington 40506, USA

L. F. M. da Silva et al. (eds.), *2nd International Conference on Advanced Joining Processes (AJP 2021)*, Proceedings in Engineering Mechanics,
https://doi.org/10.1007/978-3-030-95463-5_9

1 Introduction

It is generally accepted that steels with nitrogen content greater than 0.4 wt.% (mass percentage) in the austenitic matrix or nitrogen greater than 0.08 wt.% in the ferritic matrix, are known as high-nitrogen steels (Uggowitzer et al. 1996). In many early applications, scholars used nitrogen to replace nickel so that an austenite structure in steel could be formed (Liu et al. 2020; Vogt 2001; Kaputkina et al. 2016; Petrov et al. 1999). In the austenite, nitrogen is easier to dissolve in a solution than carbon, bringing out a less precipitation of carbides, and consequently a higher strength and corrosion resistance of steel (Qin et al. 2019; Yang et al. 2012; Franks 2015). With the development of the high-nitrogen steel, the researchers have discovered that outstanding properties of such steel, (high strength, good toughness and process performance as well as an excellent corrosion resistance) can be achieved (Li et al. 2015; Hong et al. 2011; Miura and Ogawa 2001; Thyssen and Menné 2010; Simmons 1996). By replacing nickel with nitrogen, the resulting steel features, e.g., good economy with better biocompatibility. The high-nitrogen steel has been widely used in electric power, shipbuilding, marine engineering, military equipment, medical equipment, and other fields (Stein and Hucklenbroich 2004; Kuball et al. 2020; Li et al. 2009; Nishimoto and Mori 2004).

High-nitrogen steel has been adopted as a material for structural components in various equipment structures, requiring a high load-bearing capacity and strong impact resistance. Correspondingly, the fusion welding methods such as laser welding and arc welding have been often used (Zhang et al. 2016; Zhao et al. 2007; Dong et al. 2005; Mohammed et al. 2017). However, in the fields of medical equipment applications, high-nitrogen steel has been used because the load-bearing capacity and impact resistance are not the primary factors to be considered. On the contrary, the corrosion resistance should be put forward as the priority (Baba and Katada 2006; Galloway et al.2011; Bayoumi and Ghanem 2005). Temperature in the traditional fusion welding is quite near the melting point of the high-nitrogen steel. Nitrogen in a supersaturated solid solution of the austenite has a high risk of N_2 gas formation, which greatly reduces the mechanical performance and corrosion resistance of the joint (Bang et al. 2013; Kuwana 1990; Li et al. 2011). Moreover, the nitrogen gas, which cannot escape in time will form gas voids in the welded joint, thus weaken the mechanical integrity (Geng et al. 2016; Li et al. 2019; Mohammed et al. 2016). Therefore, it is very important to keep the brazing temperature below the melt point of the high-nitrogen steel so that the loss of nitrogen could be avoided (Wang et al. 2021; Zhang et al. 2021; Shi et al. 2018).

This paper studies the brazing process of high-nitrogen austenitic stainless steel. The ultimate objective is to analyze the performance of formed joints. Furthermore, it is important to find the solution for avoiding precipitation of nitrogen during the fusion bonding. The precipitation in the second phase during the brazing process has been considered by using JMAT Pro for thermodynamic calculations. Also, the brazing holding time and brazing temperature impacts have been analyzed. The influence of joint morphology and technological parameters on the structure and mechanical properties of brazed joints have been studied. Moreover, LAMMPS software was used to conduct the molecular dynamics simulation (Swope et al. 1982; Plimpton 1995; Del Rio et al. 2011; Beeman 1976; Choudhury et al. 2011; Jawalkar et al. 2007) of the high-nitrogen steel matrix elements, such as Fe, Cu and Ni elements. The atomic diffusion process in both the Fe–Cu binary system and the Fe–Ni binary system were simulated to establish whether the intermetallic compounds would be generated. The atomic potential energy was analyzed to obtain the mean square displacement (MSD) and diffusion coefficient to compare the element diffusion ability in the Fe–Cu and Fe–Ni systems.

2 Material and Methods

2.1 The Substrate and Filler Metal

The substrate (base) metal considered is a high-nitrogen austenitic stainless steel with a high nitrogen content of 0.75 wt.%. As shown in Fig. 1, the structure of the base metal is in the forged state of equiaxial grains. The grains are mostly the austenite of uniform sizes including some crystal twinning. The base metal composition of high-nitrogen steel is as shown in Table 1. The filler adopted is Ag–Cu–Ni, in which

Fig. 1 The micro structure of the high-nitrogen austenitic stainless steel

Table 1 The base metal composition of high-nitrogen steel

Element	Cr	Ni	Mo	Mn	Cu	N	C	Fe
wt.%	24	22	8	3	0.5	0.5	≤0.02	Bal

Table 2 The chemical composition of the Ag–Cu–Ni filler

Element	Ag	Cu	Ni
wt.%	71.5	27.75	0.75

0.75% of nickel is added (silver-copper eutectic filler with a particle size of 200 mesh). The chemical composition of the filler is as shown in Table 2.

2.2 *Test Equipment and Methods*

The M60 vacuum diffusion furnace is adopted as the equipment for the vacuum brazing test. The maximum heating temperature of the vacuum furnace can reach 2150 °C, the effective space of the hot zone is Φ75 mm × 130 mm and the maximum vacuum 6×10^{-4} Pa. The AgCuNi filler was mixed with ethyl cellulose as a binder to obtain the filler paste. The test specimens were assembled in the form shown in Fig. 2 inset. No pressure was exerted on the assembled specimen during the brazing process.

The process parameters adopted were as follows: the temperature ramp up, to 850 °C, than 950 °C, and ultimately 1050 °C with holding times of 5 min, 10 min, 20 min, respectively, Fig. 2.

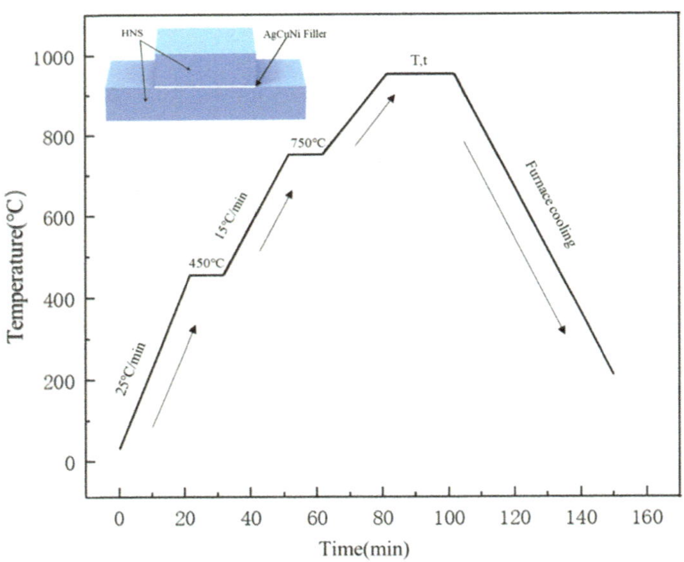

Fig. 2 Brazing process curve

2.3 Microstructure Analysis and Performance Testing

The microstructure of the joint zone was revealed by partitioning the joint with Wire EDM partitioning along the center plane of the joint, and mounted with a dental acrylic resin powder. The samples were polished by sandpaper with 400, 800, 1000, 1200, and 1500 grit, followed by 1 μm diamond polishing agent. An optical microscope was used to conduct optical characterization of the samples. The base metal needs to be corroded to uncover the morphology and the grain boundaries. We chose as an etchant the mixture of 1 ml HF, 1.5 ml HCl, 2.5 ml HNO_3 and 95 ml H_2O. Due to the high corrosion resistance of high-nitrogen steel, the corrosion time will correspondingly last as long as about 100 s. Finally, the corroded base metal and the brazed joints were observed with both metallurgical microscope and scanning electron microscope. The electronic probe was used for the microstructure analysis. The optical microscope (LEICA, DM4500P) and the scanning electron microscope (SEM, Helios Nanolab600i, FEI Company, USA) were used to analyze the microstructure and morphology of the brazed joint interfaces. The energy spectrometers were used for composition analysis.

The shear testing at room temperature has been adopted to evaluate mechanical properties response. The dimensions of the substrates are 10 mm × 10 mm × 4 mm and 10 mm × 20 mm × 4 mm. The schematic diagram of the shear strength test assembly is shown in Fig. 3. The brazed joint assembly was put into the shear fixture and tested at room temperature, with the shear test speed of 0.5 mm/s. The stress–strain data curve, and the maximum load when the joint was broken were collected during the shear test. The 3 shear tests, under each set of brazing process parameters were collected, and the average value is taken as the average joint strength to ensure the accuracy. The shear strength was measured by a universal tensile testing machine (Instron 5569, Instron Co., USA).

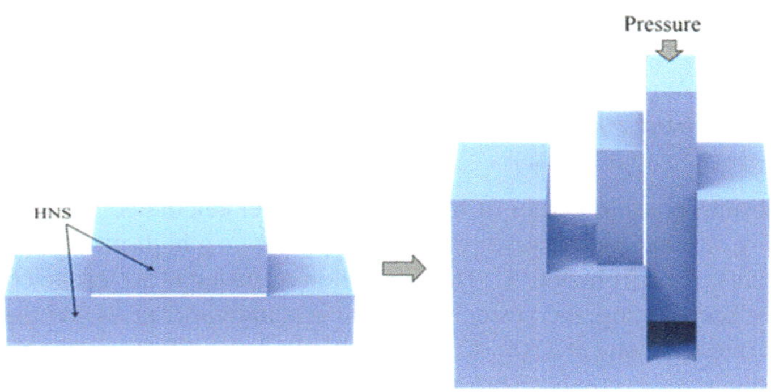

Fig. 3 The schematic diagram of the shear strength test

2.4 Molecular Dynamics Simulation on Element Diffusion Process of Filler and Base Metal

The atomic diffusion simulation of Fe–Cu binary system has been performed by using the NPT ensemble (Berendsen et al. 1984; Martyna et al. 1992; Nosé 1984; Mark et al. 2011; Gervilla et al. 2020). It is assumed that the system is at a constant temperature and pressure. The particle numbers in the system at different brazing temperatures of 850, 950 and 1050 °C have been considered as being constant. For the atomic diffusion simulation in Fe–Ni binary system, we selected the temperatures of 1123, 1223 and 1323 K. No additional pressure is needed during the brazing testing, so a standard atmospheric pressure (0.1 MPa) has been assumed. The diffusion model was constructed with crystalline Fe and crystalline Cu. The lattice constant of Cu atom is 3.61492505 nm and the lattice constant of Fe atom is 2.85532463 nm. For the molecular dynamics model of Fe–Ni diffusion, the lattice constant of Ni atom is 3.506486 nm while the lattice constant of Fe atom is as already adopted, i.e., 2.85532463 nm. For the sake of higher calculation efficiency and appropriately less calculation time, we established the supercells with a smaller number of atoms. About 10 reduplicate unit cells were adopted in the x, y, and z directions, respectively, so as to establish the supercell and keep a small distance between the two crystals. The timestep of the molecular dynamics simulations was 60 fs. The periodic boundary conditions were applied in three directions, and the EAM potential function developed by Bonny et al. (2009, 2013), was adopted to calculate the interatomic force between the Fe and Cu atoms. The potential function including the Fe–Cu–Ni atomic potential was developed by Bonny for the microstructure evolution and mechanical property changes in the steel of reactor pressure vessel as well as the atomic scale diffusion simulation.

3 Results and Discussion

3.1 The Second Phase Precipitation Behavior

The equilibrium phase diagram of high-nitrogen steel calculated by JMAT PRO is shown in Fig. 4. One can observe precipitated phases in a high-nitrogen steel when the temperature rises from 500 °C. The proportion of austenite is close to 60%. As the temperature keeps rising, the austenite content gradually increases and the structure transforms into austenite at 1200 °C, until it's melted at 1380 °C. In the process, the austenite becomes extremely stable without any high temperature δ-ferrite produced. This is mainly because of the high nitrogen and nickel in the high-nitrogen steel. The two elements can stabilize and expand the austenite phase region. This reduces the production of δ-ferrite in the high temperature region and inhibits the martensite phase transformation in the low temperature region.

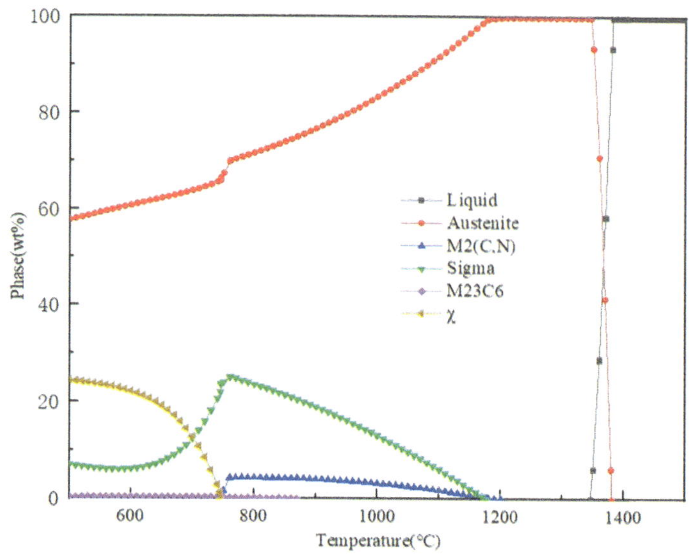

Fig. 4 The equilibrium phase diagram of high-nitrogen steel

In the considered process, the second phase that precipitates in the high-nitrogen steel mainly include M2(C,N) phase, σ phase, M23C6 phase and χ phase. The M2(C,N) phase is mainly the Cr2(C,N) phase, whose precipitation temperature range is about 750–1200 °C; the actual composition of σ phase is (Fe,Ni)x(Cr,Mo)y Its precipitation temperature range is relatively large, between 500 and 1200 °C. The peak of precipitation is at 750 °C. The M23C6 phase is mainly the Cr carbide, which has only a small amount of precipitation in the lower temperature range. The chemical composition of χ phase (Chi phase) is mostly (Fe,Ni)36Cr18Mo4 in austenitic steels. The precipitation starts at 750 °C and the proportion of precipitated phases gradually increases as the temperature keeps reducing, which mainly takes place in the low temperature region.

The calculated isothermal cooling transition curve is shown in Fig. 5. The curve can also be regarded as the isothermal precipitation curve to analyze the precipitation time of the precipitated phase when the high-nitrogen steel is held at the corresponding temperature. One can see that the σ phase is the easiest to get precipitated when the heated high-nitrogen steel is cooled, followed by the χ phase and the M2(C,N) phase. For the σ phase, the most sensitive temperature is 780 °C and the incubation period is 4.1 h. For the χ phase, the most sensitive temperature is 750 °C and the incubation period is 6.1 h. For the M2(C,N) phase, the most sensitive temperature is 750 °C and the incubation period is as long as more than 30 h.

Based on the performed analysis, one can assess the sensitive temperature of the second phase precipitation would be 700–800 °C. Therefore, with regard to the brazing process of interest in this paper, the high-nitrogen steel shall not be tested at 700-800 °C.

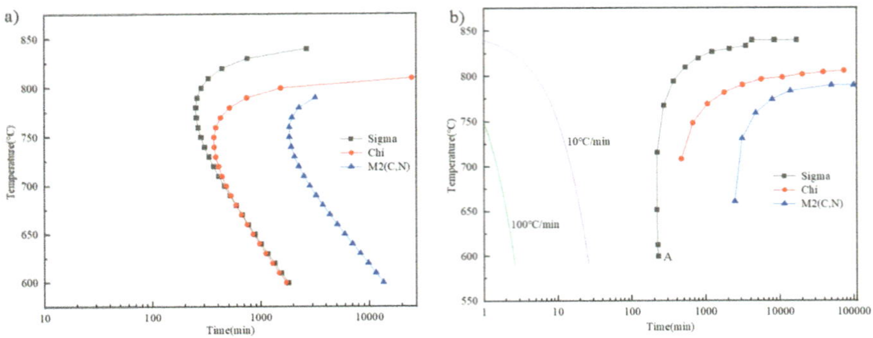

Fig. 5 High nitrogen stainless steel cooling transition curve. **a** Isothermal cooling transition **b** continuous cooling transition

The brazing temperature curve shown in Fig. 2, indicates dwell times of 30 min at 850, 950 and 1050 °C, followed by a convective cool down within the furnace. Polishing and etching followed the tests. The metallographic imaging with the Image-Pro Plus software has been performed to determine the ratio of the precipitated phase. The microstructure result is shown in Fig. 6. It can be seen that continuous precipitates appear at the grain boundaries when the brazing process curve is held at 850 °C for

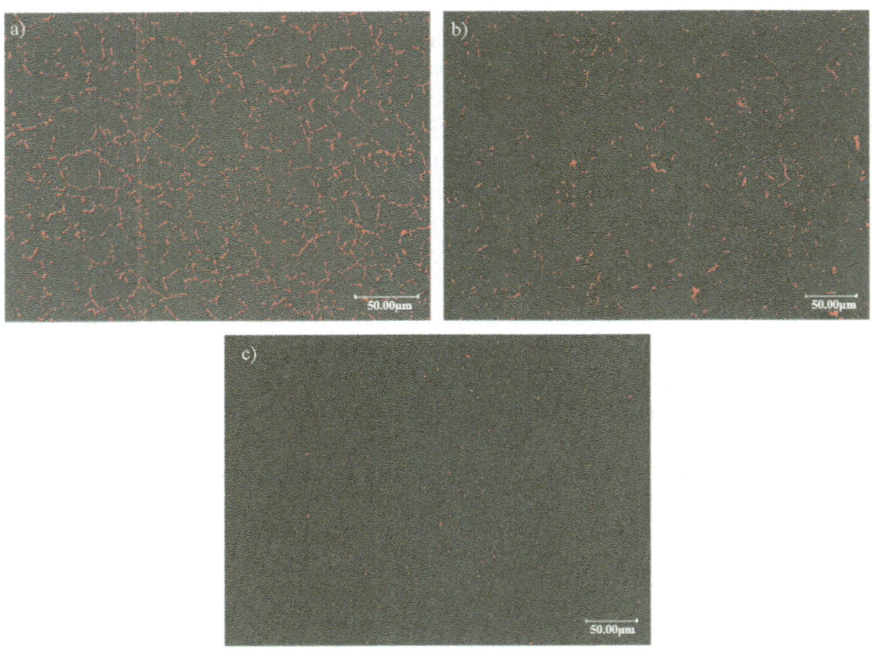

Fig. 6 Microstructure of the joints after the dwell for 30 min. **a** 850 °C, **b** 950 °C, **c** 1050 °C

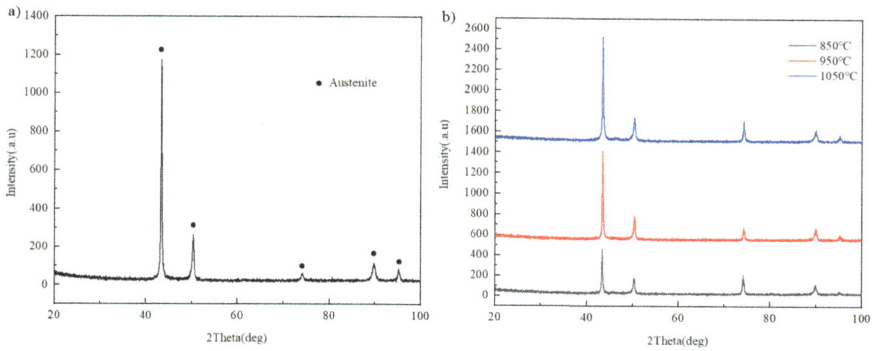

Fig. 7 XRD spectrum line of high-nitrogen steel. **a** Heat dwell at 850 °C, **b** comparison of the three heat treatments

30 min. A small amount of larger granular precipitates appear at the triple grain boundaries. At the holding temperature of 950 °C, a small amount of discontinuous precipitated phases appear on the interface. When the temperature rises to 1050 °C, almost no precipitated phases can be seen within the microstructure.

As the dwell temperature increases, the precipitates in the high-nitrogen steel gradually decrease, which conforms to the result obtained by the thermodynamic calculations. This is because at the higher temperature the deviation from the critical temperature is greater. Since the holding time of brazing is less than 30 min, much shorter than the precipitation incubation time calculated by the phase diagram, less precipitated phase can be observed in the microstructure.

To determine the precipitated phases, XRD analysis is performed. The analysis is performed on the base metal after the aforesaid three heat treatments. The XRD results of base metal after dwell at 850 °C are shown in Fig. 7. By comparing to standard PDF card with Jade software, we find that the diffraction angle is consistent with the diffraction features of austenite standard card, without the diffraction peaks of σ phase and nitride M2(C,N) phase. The comparison of the three base metals at 850, 950 and 1050 °C, one finds the same diffraction angle for the diffraction peak intensity. Accordingly, we can conclude that the base metals after brazing heating and dwell are still the austenitic structures. There are fewer precipitates, hence the phases with precipitates are not detected.

The brazing dwell and cooling process have little effect on the base metals, consequently such treatments will not produce many precipitates. Therefore, the filler AgCuNi can be used for brazing tests with the temperature range set between 850 and 1050 °C.

3.2 Microstructure Features of High-Nitrogen Steel Brazed Joint

Analysis is carried out of a typical AgCuNi high-nitrogen steel brazing joint. Under the conditions of the brazing temperature of 850 °C with 20 min of holding time, the obtained microstructure of the joint is shown in Fig. 8. It can be seen that the structure of the joint interface appears to be good, without any defects such as voids and cracks. The joint mainly consists of the Ag–Cu filler layer and the diffusion area beyond the interface between the filler and the base metal. The filler layer is mainly divided into the light-colored area A and the dark-colored area B, according to the contrast difference. The diffusion area mainly includes the high-nitrogen steel base metal and the zone resulting from the filler diffusion into the base metal.

In Fig. 9, it can be seen that the main elements in the filler metal area are Ag, Cu and Ni. These are mostly in the light color area, Ag and a few Cu and Ni, zone A. The B part of the dark regions is mostly Cu, while Ag and Ni constituents are a few. On the base material side of the high nitrogen steel, it is mainly Fe, Cr, Mn and other high nitrogen steel elements.

To determine the reaction products and element diffusion between the filler and the base metal during the brazing process, the interface between the high-nitrogen steel base metal and the filler at brazed joint was highly magnified to obtain the microstructure shown in Fig. 10. Energy Dispersive Spectrometer (EDS) analysis was done on each point shown in the figure and the corresponding chemical composition is shown in Table 3.

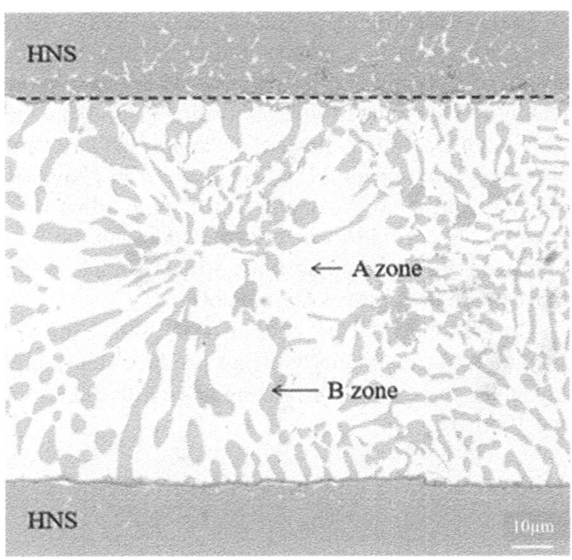

Fig. 8 The microstructure of a typical high-nitrogen steel/AgCuNi/ high-nitrogen steel brazed joint, under the brazing temperature of 850 °C and holding time of 20 min

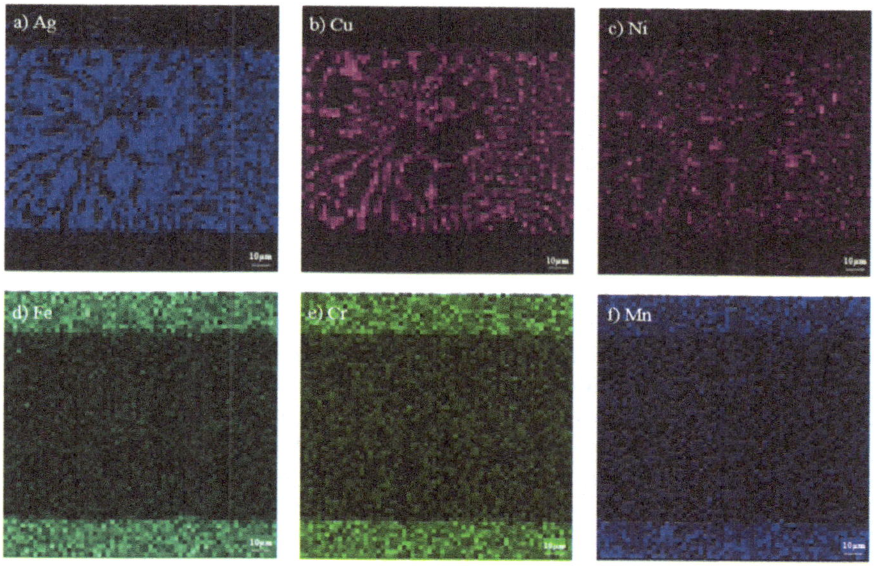

Fig. 9 Surface scan of the main elements on the joint interface. **a** Ag; **b** Cu; **c** Ni; **d** Fe; **e** Cr; **f** Mn

Fig. 10 High power microstructure of brazing joint interface

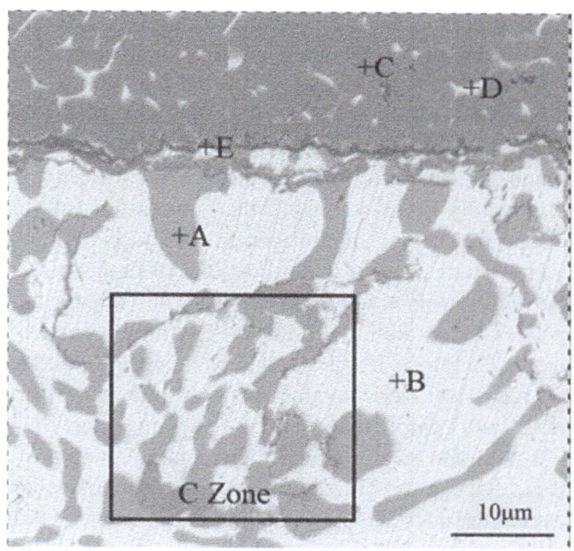

Table 3 Results of energy spectrum analysis (at.%)

	Ag	Cu	Ni	Fe	Cr	Mn	N	Possible phase
A	5.86	73.37	0.96	8.45	3.88	7.47	0	Cu[s,s]
B	68.61	14.13	0.25	3.78	4.74	8.49	0	Ag[s,s]
C	1.93	1.97	1.89	57.77	19.31	12.42	4.71	HNS
D	50.41	5.79	0.82	20.31	8.15	9.72	4.79	Ag/Cu/HNS
E	22.03	6.38	8.03	15.80	6.31	4.73	36.72	AgCu/Ni/M_2N

It can be seen from Table 3 that point A and point B in Fig. 10 are Cu-rich phase and Ag-rich phase, respectively. The element at point A is mainly Cu with relatively small amounts of Ag, Fe, Mn and other elements. That means that a small amount of Fe and alloying elements in the base metal diffuse into the filler. Since Cu can dissolve a small amount of alloying elements, point A in the dark-colored area is devised to be the Cu-based solid solution containing a small amount of Ag, Fe and Mn, so it's marked as Cu[s, s]. The element at point B is mainly Ag with a relatively small amount of Cu, Mn and other elements. Point B in the light-colored area is devised to be the Ag-based solid solution containing a relatively small amount of Cu and Mn, so it's marked as Ag[s, s]. The area C shown in Fig. 10 features as the main structure in the middle of the Ag–Cu filler layer. According to the silver-copper phase diagram, it can be seen that the filler composition adopted in this experiment is at the silver-copper eutectic point, so this area is regarded as the Ag–Cu eutectic region. In the filler layer we can find that Mn element concentration has increased compared with other elements of the base metal. That is mainly because the Mn element will dissolve in the AgCuNi filler to increase the wettability and fluidity, which contributes to further wetting and spreading of the filler. The element at point C is mainly Fe as well as a small amount of Cr-Mn alloying element and N. The atomic percentage of nitrogen is 4.71% and the mass percentage is 1.21%. The overall element ratio is close to that of high-nitrogen steel, so this area is mainly the high-nitrogen steel base metal. The element at point D is mainly Ag as well as Fe, Cr, Mn and other elements with the composition ratio close to that of base metal. Therefore, it is devised to be the area formed by the diffusion of filler in high-nitrogen steel base metal. Point E is at the interface between filler layer and base metal with complicated element structure. Apart from Ag, Cu and other elements of high-nitrogen steel, there are also a large amount of N element and an enrichment of Ni. During the brazing process, Ni in the Ni-containing filler forms a transition layer at the interface between filler and base metal thus making a better bonding with an effectively improved corrosion resistance (Baskes and Daw 1984; Chang et al. 2020). With regard to a large amount of N, it is supposed that during the brazing and heat-preservation process, N in the high-nitrogen steel base metal has stronger diffusivity, while diffusing along the direction of grain boundaries and the interface. After being aggregated at the interface, it precipitates in the form of nitrides with Fe, Cr, Mn and other elements.

To sum up, in the structure of high-nitrogen steel brazing joint, the Ag and Cu will hardly react with the alloy elements in the high-nitrogen steel without producing a new intermetallic compound reaction layer. However, the Ag–Cu filler and high-nitrogen steel base metal diffuse with each other to form the diffusion area at the joint. A small amount of Cu, Mn and other elements are dissolved in Ag in the middle of the joint zone to form the Ag-based solid solution, in which the Ag–Cu eutectic structure is distributed. The Ni, N element enrichment areas are at the interface of base metal and filler. Ag and Cu elements diffuse on the base metal side to form a white striped mixed area. Therefore, the typical joints can be divided into the zones as follows: (1) the filler layer, that is, the Ag-based solid solution and the Ag–Cu eutectic structure in the middle of the joint; (2) Ni, N element enrichment area at the transition layer of the interface between base metal and filler; (3) diffusion area on the side of high-nitrogen steel base metal.

3.3 Molecular Dynamics Analysis

3.3.1 Fe-Cu and Fe–Ni Diffusion Process

The simulation shown in Fig. 11 is the result of the calculation of Fe-Cu atom diffusion model at the times of 0 ps, 200 ps, 400 ps and 600 ps at the temperature of 1123 K (850 °C). One can see that when the diffusion gets started, the gap between Fe and Cu atoms disappears. The right side of atom distribution diagram shows the concentration distributions of Fe and Cu atoms along the Z axis in the diffusion direction, (the interface is located at 30 nm on the Z axis). By comparing the four atomic diffusion models and atomic concentration distribution diagrams at different periods of time, we can see that as the time goes by, the thickness of diffusion area gradually increases and the Fe atoms start to diffuse into the Cu lattice. The Fe atoms on the side of the Cu lattice start to increase, although only a small number of Cu atoms diffuse into the Fe lattice. During the diffusion process, the thickness of diffusion area increases as the diffusion time goes by. Note that the concentration curve of the Fe–Cu diffusion system is basically the linear one, which means that species diffuse in the holding process mutually without a mesophase compound generated. Such a result conforms to the ultimate mutual solubility in Fe–Cu binary phase diagram without the mesophase generated.

The diagram shown in Fig. 12 is the model of Fe–Ni atom diffusion at the time instances of 0 ps, 200 ps, 400 ps and 600 ps. The temperatures was 1123 K (850 °C). We can see that when the diffusion gets started, the gap between Fe and Ni atoms disappears immediately. The subsequent diffusion and migration processes can be observed. Similar to the Fe–Cu diffusion, the right side, see Fig. 12, shows the atom concentration distribution along the diffusion direction The interface is located at the distance of 30 nm. The thickness of diffusion area gradually increases as the diffusion time evolves. The Fe–Ni atom diffusion is also known as the asymmetric. More Fe atoms diffuse into the Ni lattice while only a small number of Ni atoms

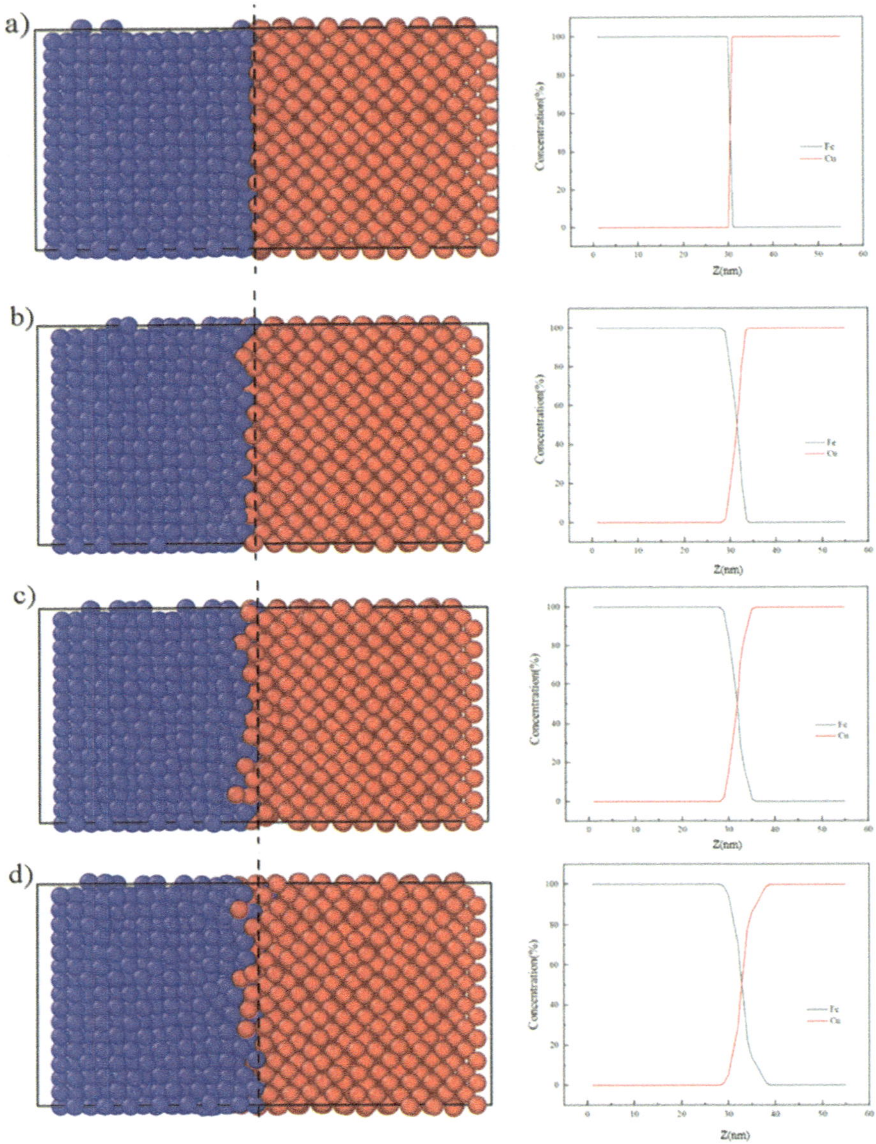

Fig. 11 The atomic distribution diagram and the concentration diagram along Z axis of Fe–Cu for different diffusion time (850 °C): **a** 0 ps; **b** 200 ps; **c** 400 ps; **d** 600 ps

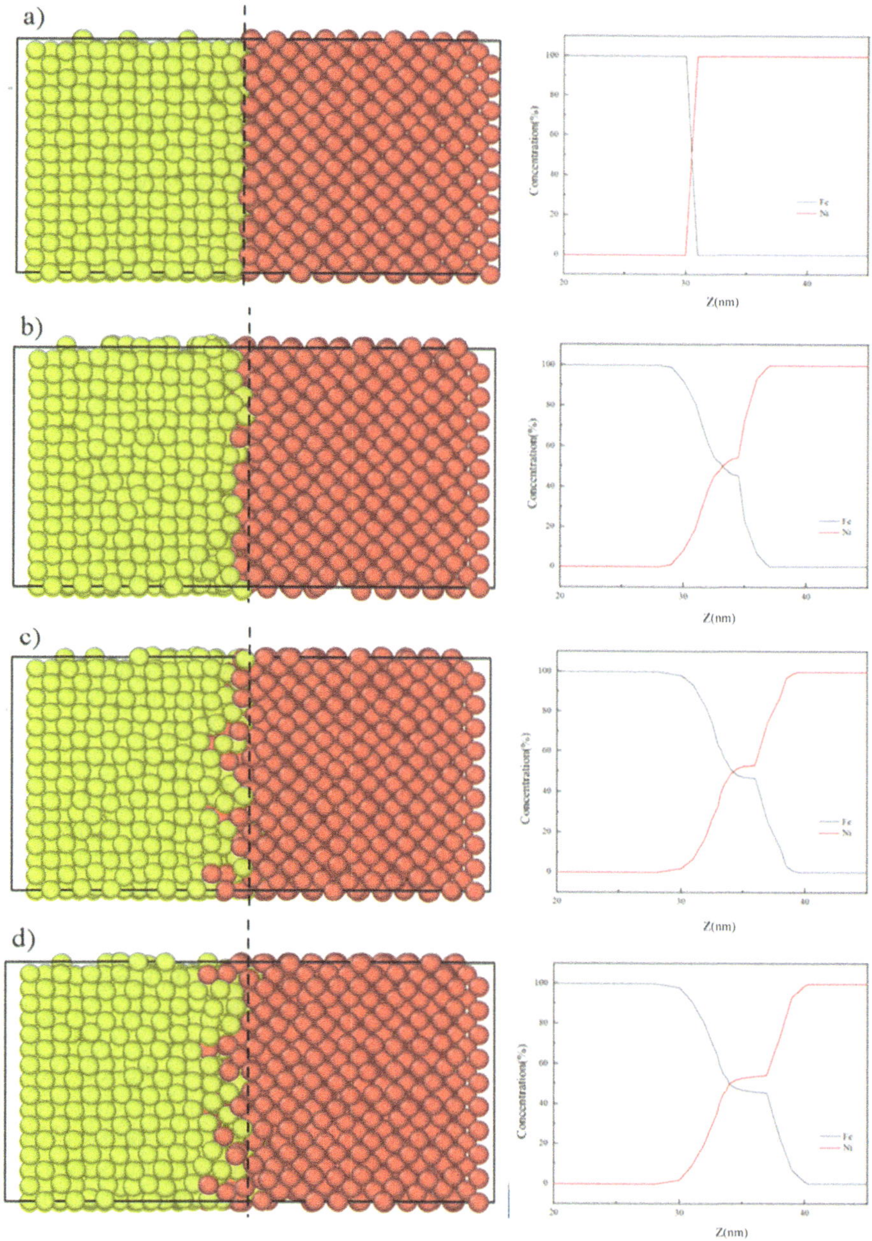

Fig. 12 The atomic distribution diagram and concentration diagram along Z axis of Fe–Ni with different diffusion time (850°C): **a** 0 ps; **b** 200 ps; **c** 400 ps; **d** 600 ps

diffuse into the Fe lattice. It can be inferred that the diffusion ability of Fe atoms is stronger than that of Ni atoms in the Fe–Ni binary system.

Compared to the diffusion in the Fe–Cu binary system, the diffusion area of Fe–Ni system is thicker, with more obvious diffusion under the same time and temperature conditions. Moreover, compared to the Fe–Cu diffusion system, the concentration curve of Fe–Ni system shows a section with a smaller slope in the diffusion domain. A mesophase is formed in this domain during the diffusion process, which means a certain amount of Fe–Ni compound is generated (Ouyang et al. 2018). By referring to the Fe–Ni binary phase diagram, it can be seen that some mesophase including FeNi, and $FeNi_3$ had been generated between Fe and Ni. This is consistent with the conclusion of the molecular dynamic simulation.

3.3.2 Atomic Potential Energy in Diffusion Process

At the diffusion time of 600 ps and at three different temperatures, we have compared the atomic potential energy in the considered Fe–Cu diffusion process. The result is shown in Fig. 13. The potential energy of an atom is zero in the equilibrium position. When the atom is away from the equilibrium position, the potential energy changes to a negative value. The smaller distance between atoms, the greater deviation from the equilibrium position and the greater absolute value of atomic potential energy. It can be seen from the figure that in the diffusion process, the absolute potential energy value of Fe atoms is greater than that of Cu atoms. This means that the Fe atoms are further away from the equilibrium position and thus become more unstable. By averaging the atomic potential energy of Fe and Cu at the considered three temperatures, one gets the results are shown in Table 4. For the same Fe or Cu atoms, the absolute value of the atomic potential energy increases as the temperature increases, and the atoms become more unstable and easier to break the barrier to get diffused.

Similarly, we have calculated the potential energy of each atom in the diffusion process based on a molecular dynamics simulation. At the diffusion time of 600 ps and at the three temperatures, we have compared the atomic potential energy in Fe–Ni diffusion process. The result is shown in Fig. 14. It can be seen that during the diffusion process, the absolute potential energy value of Fe atoms is smaller than that of Ni atoms. This means that the atoms start to move away from the equilibrium positions. Moreover, the Ni atoms get further away from the equilibrium position and become more unstable. By averaging the atomic potential energy of Fe and Ni at three temperatures, one obtains the results are shown in Table 5. For the same Fe or Ni atoms, the result of atomic potential energy appears to be the same as that obtained in the Fe–Cu diffusion simulation. As the temperature increases, the absolute value of atomic potential energy increases and the atoms become more unstable and easier to break the barrier, hence to get diffused.

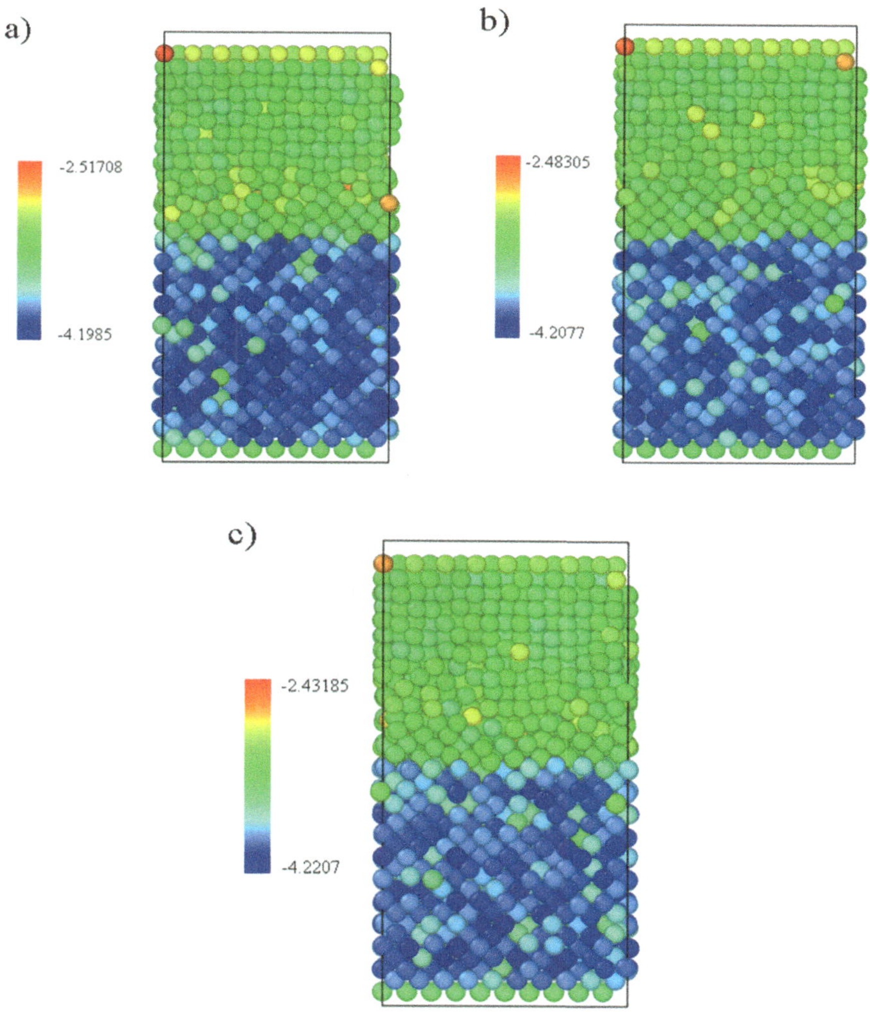

Fig. 13 Diagram of Fe–Cu atomic potential energy at 600 ps and the three different temperatures; **a** 1123 K; **b** 1223 K; **c** 1323 K

Table 4 Average potential energy of Fe and Cu atoms at different temperatures

Temperature	1123 K	1223 K	1323 K
Fe	−3.8522 eV	−3.8717 eV	−3.8829 eV
Cu	−3.2708 eV	−3.2881 eV	−3.2975 eV

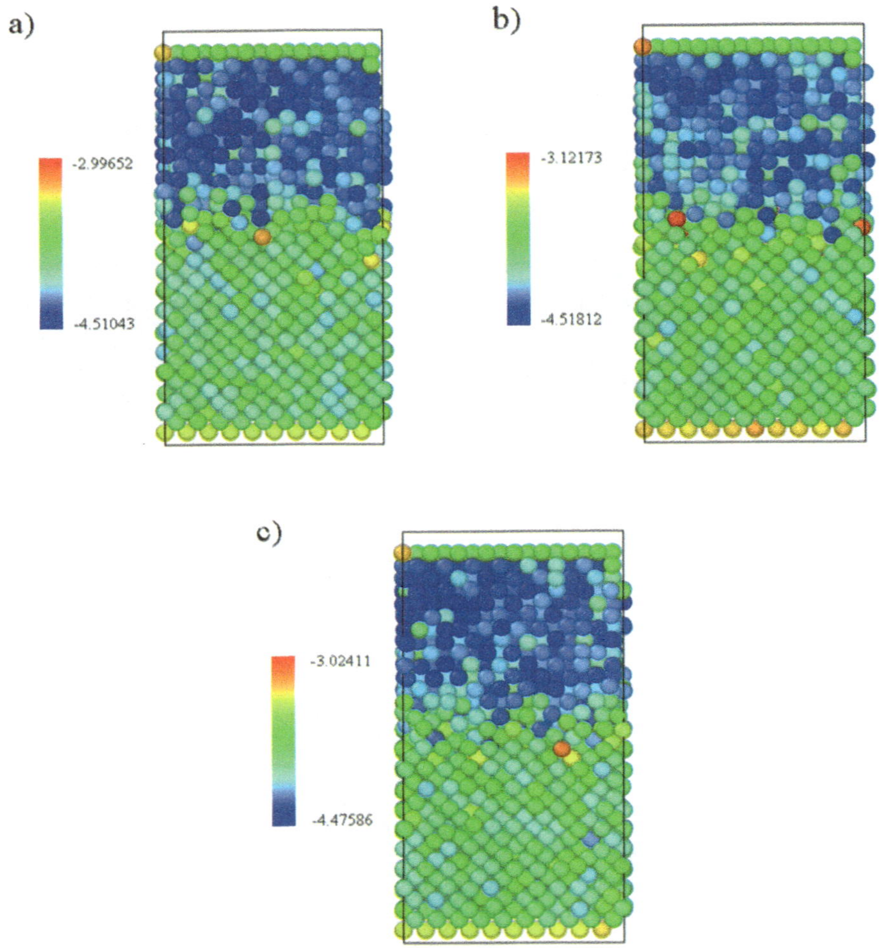

Fig. 14 Diagram of Fe–Ni atomic potential energy at 600 ps in different temperatures; **a** 1123 K, **b** 1223 K, **c** 1323 K

Table 5 Average potential energy of Fe and Ni atoms at different temperatures

Temperature	1123 K	1223 K	1323 K
Fe	−3.8532 eV	−3.8731 eV	−3.8847 eV
Ni	−4.1838 eV	−4.1985 eV	−4.2066 eV

3.3.3 Diffusion Coefficient

The Mean Square Displacement (MSD) evaluations indicate different diffusion ability of Fe and Cu atoms as well as the Fe and N atoms, Tables 6 and 7. These data

Table 6 Comparison of the MSD of Fe and Cu in different time at 1323 K

Time (ps)	60	180	300	480	600
MSD of Fe ($10^{-9}m^2$)	3.27739	5.33243	7.77702	11.4883	14.9785
MSD of Cu ($10^{-9}m^2$)	2.56587	4.73497	7.03857	10.8754	13.7495

Table 7 Comparison of the MSD of Fe and Ni in different time at 1323 K

Time (ps)	60	180	300	480	600
MSD of Fe ($10^{-9}m^2$)	2.60847	3.16528	3.87163	5.36935	6.3628
MSD of Ni ($10^{-9}m^2$)	2.46461	4.73497	3.64864	5.04519	5.9719

can be used to calculate the diffusion coefficient.

The MSD comparison of Fe and Cu atoms is shown in Table 6, from which we can see that at the given temperature, the MSD of Fe atoms is generally higher than that of Cu atoms. Therefore, we can infer that the diffusion ability of Fe atoms is greater than that of Cu atoms in the Fe-Cu binary diffusion system.

Similarly, the MSD of Fe atoms and Ni atoms increases linearly as the diffusion process continues. The MSD comparison of Fe and Ni atoms is shown in Table 7, from which we can see that at the given temperature, the MSD of Fe atoms is generally higher than that of Ni atoms. Therefore, we can infer that the diffusion ability of Fe atoms is greater than that of Ni atoms in the Fe–Ni binary diffusion system.

4 Conclusions

In this study, high nitrogen steel was bonded by a vacuum brazing process with AgCuNi as a filler braze. The second phase precipitation of high nitrogen steel under brazing process was empirically and numerically considered. Typical interface morphology of the joints was analyzed. Moreover, we have discussed the influence of different technological parameters on the joint interface structure and mechanical properties. The specific results of this study can be summarized as follows:

(1) The second phase precipitation behavior of high nitrogen steel exposed to a brazing process has been analyzed. The phase diagram and cooling transition curve of high nitrogen steel have been obtained through a thermodynamic calculation. A conclusion has been reached that the critical temperature range for the second phase precipitates is 700°C-800°C. Therefore, the brazing process should be avoided in that temperature range. The calculation illustrates that the cooling process has little significance for the cooling speed, and the second phase can be avoided during the furnace. We have established that the heat treatment of the brazing process studied had a little influence on the

high nitrogen steel. The brazing temperature range was determined to be 850 °C–1050 °C.

(2) The brazing heating and cooling treatment impact on the high nitrogen steel was studied. Furthermore, the precipitation of the second phase in the high nitrogen steel has been considered. Because the critical temperature range is avoided and the brazing holding time is shorter than the "incubation period" of the precipitated phase, so the precipitation is reduced. The composition of the precipitated phase cannot be determined by XRD.

(3) The brazing process is dominantly characterized by diffusion between the migrating elements. Typical joint structure is mainly the Ag–Cu eutectic of the filler layer with the element diffusion domain on the substrate material side. Typical joints domains can be divided into the areas as follows: (1) the filler layer, that is, the Ag-based solid solution and the Ag–Cu eutectic structure in the middle of the resolidified braze; (2) Ni and N enrichment area within the transition layer of the interface between the substrate (base) metal and filler; (3) elements diffusion area on the side of the high-nitrogen steel base metal.

(4) A simulation of molecular dynamics diffusion was conducted on the Fe-Cu and Fe–Ni binary systems respectively and obvious mutual diffusion phenomena were found. The thickness of the diffusion layer has increased with the diffusion time. However, according to the distribution curve of atomic concentration, one can see that there is only a mutual diffusion process in the Fe-Cu binary system while in the Fe–Ni binary system there are both a diffusion process and a generation of the Fe–Ni compound mesophase. According to the calculation and analysis of (i) the element diffusion MSD and (ii) the diffusion coefficient in the two systems at different temperatures, one can see that the higher temperature, the greater is atom MSD and the diffusion coefficient, with a stronger diffusion capacity.

Acknowledgements The authors acknowledge the Natural Science Foundation Excellent Youth Project of Henan (Grant number: 202300410268), and the China Postdoctoral Science Foundation (Project number: 2019M651280, 2019M662011), Open Fund of State Key Laboratory of Advanced Brazing Filler Metals and Technology (Project number: SKLABFMT201901).

Conflicts of Interest

The authors declare that they have no conflict of interest.

Data Availability Statement

The raw/processed data required to reproduce these findings cannot be shared at this time due to technical limitations.

References

Baba, H., Katada, Y.: Effect of nitrogen on crevice corrosion in austenitic stainless steel. Corros. Sci. **48**(9), 2510–2524 (2006)

Bang, K., Pak, S., Ahn, S.: Evaluation of weld metal hot cracking susceptibility in superaustenitic stainless steel. Met. Mater. Int. **19**(6), 1267–1273 (2013)
Baskes, M.I., Daw, M.S.: Embedded-atom method: Derivation and application to impurities, surfaces, and other defects in metals. Phys. Rev. B **29**(12), 6443–6453 (1984)
Bayoumi, F.M., Ghanem, W.A.: Effect of nitrogen on the corrosion behavior of austenitic stainless steel in chloride solutions. Mater. Lett. **59**(26), 3311–3314 (2005)
Beeman, D.: Some multistep methods for use in molecular dynamics calculations. J. Comput. Phys. **20**(2), 130–139 (1976)
Berendsen, H.J.C., Postma, J.P.M., Van Gunsteren, W.F., DiNola, A., Haak, J.R.: Molecular dynamics with coupling to an external bath. J. Chem. Phys. **81**(8), 3684–3690 (1984)
Bonny, G., Pasianot, R.C., Castin, N., Malerba, L.: Ternary Fe–Cu–Ni many-body potential to model reactor pressure vessel steels: first validation by simulated thermal annealing. Philos. Mag. **89**(34–36), 3531–3546 (2009)
Bonny, G., Castin, N., Terentyev, D.: Interatomic potential for studying ageing under irradiation in stainless steels: the FeNiCr model alloy. Model Simul. Mater. Sci. **21**(8), 85004 (2013)
Choudhury, S., Barnard, L., Tucker, J.D., Allen, T.R., Wirth, B.D., Asta, M., Morgan, D.: Ab-initio based modeling of diffusion in dilute bcc Fe–Ni and Fe–Cr alloys and implications for radiation induced segregation. J. Nucl. Mater. **411**(1), 1–14 (2011)
Del Rio, E., Sampedro, J.M., Dogo, H., Caturla, M.J., Caro, M., Caro, A., Perlado, J.M.: Formation energy of vacancies in FeCr alloys: dependence on Cr concentration. J. Nucl. Mater. **408**(1), 18–24 (2011)
Dong, W., Kokawa, H., Sato, Y.S., Tsukamoto, S.: Nitrogen desorption by high-nitrogen steel weld metal during CO_2 laser welding. Metall. Mater. Trans. B **36**(5), 677–681 (2005)
Franks, R.: Tension and compression stress-strain characteristics of cold-rolled austenitic chromium-nickel and chromium-manganese-nickel stainless steels. J. Aeronaut. Sci. **9**(11), 419–438 (2015)
Galloway, A., McPherson, N., Baker, T.: An evaluation of weld metal nitrogen retention and properties in 316LN austenitic stainless steel. Proc. Inst. Mech. Eng. Part L J. Mater. Design Appl. **225**, 61–69 (2011)
Geng, X., Feng, H., Jiang, Z., Li, H., Zhang, B., Shucai, Z., Wang, Q., Li, J.: Microstructure, mechanical and corrosion properties of friction stir welding high nitrogen martensitic stainless steel 30Cr15Mo1N. Metals Basel **6**, 301 (2016)
Gervilla, V., Zarshenas, M., Sangiovanni, D.G., Sarakinos, K.: Anomalous versus normal room-temperature diffusion of metal adatoms on graphene. J. Phys. Chem. Lett. **11**(21), 8930–8936 (2020)
Hong, C., Shi, J., Sheng, L., Cao, W., Hui, W., Dong, H.: Influence of hot working on microstructure and mechanical behavior of high nitrogen stainless steel. J. Mater. Sci. **46**(15), 5097–5103 (2011)
Jawalkar, S.S., Raju Halligudi, S.B., Sairam, M., Aminabhavi, T.M.: Molecular modeling simulations to predict compatibility of poly (vinyl alcohol) and chitosan blends: A comparison with experiments. J. Phys. Chem. B **111**(10), 2431–2439 (2007)
Jinbao, C., Yingli, Z., Shuang, J.: Effect of solution treatment and aging on microstructure of low-nickel high nitrogen austenitic stainless steel. Heat Treat. Met. **45**(02), 120–124 (2020)
Kaputkina, L.M., Svyazhin, A.G., Smarygina, I.V., Bobkov, T.V.: Corrosion resistance of high-strength austenitic chromium–nickel–manganese steel containing nitrogen. Steel Translation **46**(9), 644–650 (2016)
Kuball, C.M., Jung, R., Uhe, B., Meschut, G., Merklein, M.: Influence of the process temperature on the forming behaviour and the friction during bulk forming of high nitrogen steel. J. Adv. Joining Process. **1**, 100023 (2020)
Kuwana, T.: The Oxygen and Nitrogen Absorption of Iron Weld Metal During Arc Welding, pp. 117–128. Springer Netherlands (1990)
Li, H., Jiang, Z., Cao, Y., Zhang, Z.: Fabrication of high nitrogen austenitic stainless steels with excellent mechanical and pitting corrosion properties. Int. J. Miner. Metall. Mater. **16**, 387–392 (2009)

Li, H.B., Jiang, Z.H., Ma, Q.F., Li, W.M.: Manufacturing high nitrogen austenitic stainless steels by pressurized induction furnace. Appl. Mech. Mater. **52–54**, 1687–1691 (2011)

Li, H.B., Jiang, Z.H., Feng, H., Zhang, S.C., Li, L., Han, P.D., Misra, R.D.K., Li, J.Z.: Microstructure, mechanical and corrosion properties of friction stir welded high nitrogen nickel-free austenitic stainless steel. Mater. Design **84**, 291–299 (2015)

Li, J., Li, H., Peng, W., Xiang, T., Xu, Z., Yang, J.: Effect of simulated welding thermal cycles on microstructure and mechanical properties of coarse-grain heat-affected zone of high nitrogen austenitic stainless steel. Mater. Char. **149**, 206–217 (2019)

Liu, Z., Fan, C., Chen, C., Ming, Z., Wang, L.: Optimization of the microstructure and mechanical properties of the high nitrogen stainless steel weld by adding nitrides to the molten pool. J. Manuf. Process **49**, 355–364 (2020)

Mark, P., Zhang, Q., Czjzek, M., Brumer, H., Ågren, H.: Molecular dynamics simulations of a branched tetradecasaccharide substrate in the active site of a xyloglucan endo-transglycosylase. Mol. Simulat. **37**(12), 1001–1013 (2011)

Martyna, G.J., Klein, M.L., Tuckerman, M.: Nosé-Hoover chains: the canonical ensemble via continuous dynamics. J. Chem. Phys. **97**(4), 2635–2643 (1992)

Miura, H., Ogawa, H.: Preparation of nanocrystalline high-nitrogen stainless steel powders by mechanical alloying and their hot compaction. Mater. Trans. **42**(11), 2368–2373 (2001)

Mohammed, R., Reddy, G.M., Rao, K.S.: Effect of filler wire composition on microstructure and pitting corrosion of nickel free high nitrogen stainless steel gta welds. T. Indian I. Metals **69**(10), 1919–1927 (2016)

Mohammed, R., Madhusudhan Reddy, G., Srinivasa Rao, K.: Welding of nickel free high nitrogen stainless steel: Microstructure and mechanical properties. Defence Technol. **13**(2), 59–71 (2017)

Nishimoto, K., Mori, H.: Hot cracking susceptibility in laser weld metal of high nitrogen stainless steels. Sci. Technol. Adv. Mat. **5**(1), 231–240 (2004)

Nosé, S.: A molecular dynamics method for simulations in the canonical ensemble. Mol. Phys. **52**(2), 255–268 (1984)

Ouyang, Y., Wu, J., Zheng, M., Chen, H., Tao, X., Du, Y., Peng, Q.: An interatomic potential for simulation of defects and phase change of zirconium. Comp. Mater. Sci. **147**, 7–17 (2018)

Petrov, Y.N., Gavriljuk, V.G., Berns, H., Escher, C.: Nitrogen partitioning between matrix, grain boundaries and precipitates in high-alloyed austenitic steels. Scripta Mater. **40**(40), 669–674 (1999)

Plimpton, S.: Fast parallel algorithms for short-range molecular dynamics. J. Comput. Phys. **117**(1), 1–19 (1995)

Qin, Y., Li, J., Herbig, M.: Microstructural origin of the outstanding durability of the high nitrogen bearing steel X30CrMoN15-1. Mater. Char. **159**, 110049 (2019)

Shi, H., Yu, Z., Cho, J.: A study on the microstructure and properties of brazing joint for Cr18-Ni8 steel using a BNi7+9%Cu mixed filler metal. Vacuum **151**, 226–232 (2018)

Simmons, J.W.: Overview: high-nitrogen alloying of stainless steels. Mater. Sci. Eng., A **207**(2), 159–169 (1996)

Stein, G., Hucklenbroich, I.: Manufacturing and applications of high nitrogen steels. Mater. Manuf. Process. **19**(1), 7–17 (2004)

Swope, W.C., Andersen, H.C., Berens, P.H., Wilson, K.R.: A computer simulation method for the calculation of equilibrium constants for the formation of physical clusters of molecules: application to small water clusters. J. Chem. Phys. **76**(1), 637–649 (1982)

Thyssen, J.P., Menné, T.: Metal allergy—a review on exposures, penetration, genetics, prevalence, and clinical implications. Chem. Res. Toxicol. **23**(2), 309–318 (2010)

Uggowitzer, P., Magdowski, R., Speidel, M.: Nickel free high nitrogen austenitic steels. Isij. Int. **36**, 901–908 (1996)

Vogt, J.B.: Fatigue properties of high nitrogen steels. J. Mater. Process. Tech. **117**(3), 364–369 (2001)

Wang, Q., Zhang, S., Lin, T., Zhang, P., He, P., Paik, K.: Highly mechanical and high-temperature properties of Cu–Cu joints using citrate-coated nanosized Ag paste in air. Progress Nat. Sci. Mater. Int. **31**(1), 129–140 (2021)

Yang, K., Ren, Y., Wan, P.: High nitrogen nickel-free austenitic stainless steel: a promising coronary stent material. Sci. China Technol. Sci. **55**(2), 329–340 (2012)

Zhang, H., Wang, D., Xue, P., Wu, L.H., Ni, D.R., Ma, Z.Y.: Microstructural evolution and pitting corrosion behavior of friction stir welded joint of high nitrogen stainless steel. Mater. Design **110**, 802–810 (2016)

Zhang, S., Wang, Q., Lin, T., Zhang, P., He, P., Paik, K.: Cu–Cu joining using citrate coated ultra-small nano-silver pastes. J. Manuf. Process **62**, 546–554 (2021)

Zhao, L., Tian, Z., Peng, Y.: Porosity and nitrogen content of weld metal in laser welding of high nitrogen austenitic stainless steel. Isij Int. **47**(12), 1772–1775 (2007)

Additive Manufacturing with Borosilicate Glass and Soda-Lime Glass

F. Fröhlich, J. Hildebrand, and J. P. Bergmann

Abstract In this paper, the possibility of additively producing individual 3D-structures with borosilicate and soda-lime glass based on a laser process with glass rods is demonstrated. In a previous study, process parameters, which show the geometrically defined deposition of individual layers were determined (Fröhlich et al. in Herstellung individueller Strukturen aus silikatischen Werkstoffen mittels Wire-Laser Additive Manufacturing. Wilhelm & Sohn, pp. 287–297, 2020). These findings are used to create single-layer walls, flat applications and free-form structures. The focus is on the additive manufacturing of large-volume components compared to the current state of the art. A CO_2 laser is used to create a melt pool on the substrate surface. The rod-based additional material is melted to create 3D-structures. The ratio between process speed and feed speed, as well as the laser power, remains constant during the experiments, as does the temperature in the process chamber. The fabricated structures are subjected to thermal post-treatment to reduce thermal stress. Remaining residual stresses are investigated with photoelasticity. The samples are quantified with destructive and non-destructive materialographical tests to determine the geometric dimensions. Taking into account the previous process parameters, the results achieved are discussed and evaluated with regard to the width, height of the layer and the bonding angle. The height adjustment in relation to the layer height is discussed in order to avoid shape deviation. The major objective of the investigation is to achieve near-net-shape production with a low tensile residual stress state in order to reduce the degree of post-processing.

Keywords Additive manufacturing · Glass · Borosilicate glass · Soda-lime glass · CO_2 laser

F. Fröhlich (✉) · J. Hildebrand · J. P. Bergmann
Production Technology Group, Technische Universität Ilmenau, Gustav-Kirchhoff-Platz, 98693 Ilmenau, Germany
e-mail: fabian.froehlich@tu-ilmenau.de

J. Hildebrand
e-mail: joerg.hildebrand@tu-ilmenau.de

J. P. Bergmann
e-mail: jeanpierre.bergmann@tu-ilmenau.de

L. F. M. da Silva et al. (eds.), *2nd International Conference on Advanced Joining Processes (AJP 2021)*, Proceedings in Engineering Mechanics,
https://doi.org/10.1007/978-3-030-95463-5_10

1 Introduction

Properties of glass, such as transparency, thermal linear expansion, chemical and thermal resistance, is increasingly used in additive manufacturing technologies this material. The different technologies open up new possibilities in the production of glass and functionalities of the resulting components, which are not technically possible with the previous technologies. The layer-by-layer construction of components gives the designer freedom in functionality and design, which is limited by the starting material and the additive manufacturing technology. Possible 3D-Printing technologies for glass are Selective Laser Sintering (SLS) (Klocke et al. 2004; Lee et al. 1995), Selective Laser Melting (SLM) (Fateri et al. 2014; Fateri and Gebhardt 2015; Kinzel et al. 2014; Khmyrov et al. 2014; Khmyrov et al. 2016), Fused Deposition Modeling (FDM) (Klein et al. 2016, 2012; Seel et al. 2018; Gal-Or et al. 2019), Stereolithography (SLA) (Kotz et al. 2017, 2016; Liu et al. 2018), Direct Ink Writing (Nguyen et al. 2017) and Direct Energy Deposition-Laser (DED-L) (Fröhlich et al. 2020; Kinzel et al. 2014; Luo et al. 2016, 2017, 2018; Witzendorff et al. 2018; Grabe et al. 2021).

However, glass has extremely high melting temperatures, non-linear thermo-mechanical properties of glass and is sensitive to thermal shocks (Schaeffer and Langfeld 2020). This poses different challenges for the additive manufacturing processes mentioned. In the SLS or SLM process, glass is typically fed as a powder material. During production, the laser moves over the powder bed layer by layer and scans the object contour. Depending on the possible packing density of the glass powder, cavities between the grains being fused during production and remaining in the component lead to defects similar to bubbles (Kinzel et al. 2014). If the density of the manufactured component is to be increased, it is subjected to thermal post-treatment. This subjects the component to shrinkage and consequently to geometric deviation (Klocke et al. 2004). Kinzel et al. (2014) shows the advantages of using solid material as compared to powder to manufacture optical components. In the AM technology SLA and Direct Ink Writing, a composite material containing glass powder is used as the starting material (Kotz et al. 2017, 2016; Nguyen et al. 2017). The production of a green body from the glass-containing composite material can be done at room temperature. To obtain a glassy body, the green body compact must be subjected to a long heat treatment process. During this process, the solvents and organic binders outgas from the green body. The outgassing requires volatile binders, which also limit the thickness of the wall structures produced (Gal-Or et al. 2019). However, these processes are suitable for small-scale objects with a high level of detail. This contrasts with FDM, where Klein et al. (2016) can achieve a print volume of 460 mm^3/s according to the design. The average height of a layer was 4.5 mm and the width 9.5 mm with a contour deviation of 0.5 mm (Klein et al. 2016). A typical feature of the technology is the distinct layer structure, which creates a wavy surface. Because gravity is used to feed the molten glass from the crucible through the nozzle, a constant fill level is necessary here. The varying filling level had a negative effect on the printing result. It was observed that the glass stuck to the nozzle during printing rather than to the colder previous printed layer (Klein et al. 2016). The wear that occurs on the crucible, i.e. the detachment of surface particles from the crucible wall by the molten glass, also has a negative effect on the printing result (Vogel 1993). In the DED-L technology, a melt is created on the sample surface of the glass with the energy carrier laser, into which the filler material is fed and melted (Fröhlich et al. 2020; Kinzel et al. 2014; Luo et al. 2016, 2017, 2018; Witzendorff et al. 2018; Grabe et al. 2021). For glass processing, the use of a CO_2 laser with a wavelength of 10.6 μm is typical, as glass is opaque in this range (Gräf et al. 2013). Defects caused by inhomogeneities due to powdery starting materials or particles of the technical periphery, stress cracks due to wall thicknesses and lack of opportunities for outgassing solvents and organic binders can be avoided for large-format components if solid material is used for additive manufacturing without intermediate steps. If fibres are used as filler material, they are covered with a polymer coating for better handling. The coating is removed mechanically in an additional step before feeding or melted off by the laser (Witzendorff et al. 2018; Grabe et al. 2021). However, there is a risk that particle residues can get into the melt. For optical elements, Luo et al. (2018) have shown that the use of glass rods achieves good results. If glass rods made of soda-lime glass are used, it is necessary to preheat the substrate with a hot plate in order to keep the thermal shock as low as possible (Luo et al. 2017). Due to the high thermal shock resistance of quartz glass, a heating plate for substrate preheating is not necessary (Luo et al. 2018; Grabe et al. 2021). Increasing the number of layers, the distance between the manufacturing point and the preheated substrate will be greater. Due to the poor thermal conductivity of glass, only the first layers are sufficiently preheated for manufacturing. From the point of lowering heat influence from the hot plate, the thermal gradient between the fabrication point and the base plate increases. This leads for glasses like soda lime and borosilicate glass to a greater susceptibility to cracking during the additive manufacturing process.

In this work, the possibility of additive manufactured individual 3D structures from borosilicate and soda-lime glass based on a laser process with glass rods is demonstrated. Knowledge from previous work is used to create single-layer walls, planar applications and individual structures. The focus is on the additive manufacturing of large-volume components compared to the current state of the art. In addition, an experimental setup will be created that can process several types of glass.

2 Experimental Setup and Preliminary Work

With the aim of developing a process that guarantees the additive manufacturing of individual structures made of glass, both in height and in planar area, the greatest potential was seen in the Direct Energy Deposition Laser (DED-L) process. A CO_2 laser is used to create a melt pool on the surface of the substrate. The rod-shaped additional material is fed and melted to create 3D structures (Fig. 1) (Fröhlich et al. 2020). The relative movement in the plane (X–Y) of the specimen to the working point is implemented by two linear tables offset by 90° in their direction of movement.

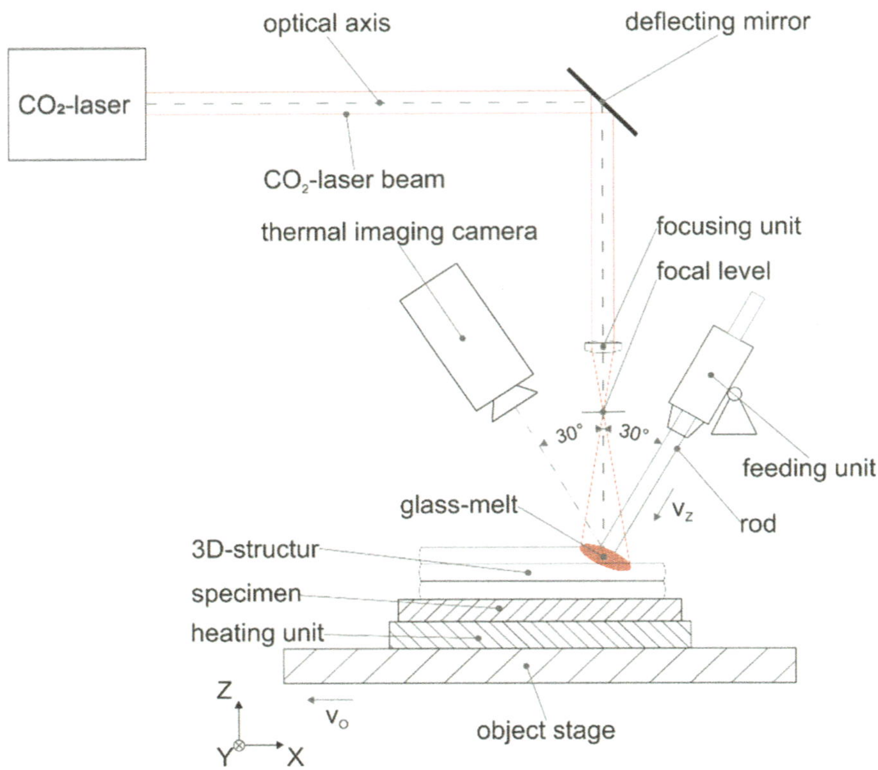

Fig. 1 Sketch of the experimental setup of the Direct Energy Deposition Laser process (Fröhlich et al. 2020)

Geared DC motors with the SM 400 control move these. The shift of the working point in height is implemented by moving the focussing unit along the Z-axis. A heating plate is used to reduce the thermal gradient in the additive manufacturing of borosilicate and soda lime glass. The accessibility of the set-up enables the use of non-contact measuring methods. A thermal imaging camera (model MC320FHT) from the company LumaSense Technologies GmbH is used to monitor the temperatures generated in the process. The calibrated temperature range extends from 400 to 1800 °C. Maximum temperatures of up to 2100 °C can be measured. The measurement uncertainty according to the manufacturer is ± 2 K or 2% of the measured value. The thermal imaging camera is aligned at an angle of 30° to the laser beam and orthogonal to the direction of manufacturing. The reduction of the thermal gradient generated by the laser beam from the glass surface during additive manufacturing is implemented by the focusing unit method. A stronger defocusing of the laser beam increases the interaction range between the laser beam and the glass surface. The CO_2 laser beam source used (Synrad firestar f400, $\lambda = 10.6$ µm, continuous wave) is aligned perpendicular to the specimen. The majority of the area of the incident laser beam diameter is used to heat the substrate. The calculated interaction diameter between laser and glass surface is approximately 10 mm. The distance between the focusing unit and the substrate surface was not changed during the experiments. The feeding unit is a proprietary development. A stepper motor constantly feeds the additional material into the melt at a speed of 1 mm/s via a linear rail. By means of a manual three-axis positioning system, the feed point of the additional material can be adjusted to the melt. The feed angle is adjusted by a turntable and remains constant during the tests at an angle of 30° between the feed axis and the CO_2 laser beam. The direction of feed and the direction of movement of the object table are in the same direction, which results in a dragging arrangement. Borofloat® and float glass from SCHOTT are used as specimens for the experiments. The dimensions of the specimens are 50 × 50 × 3 mm. The additional material used are glass rods made of Duran® (2 mm diameter) and AR-Glas® (3 mm diameter) from SCHOTT. It has been shown that the thermal gradient between the substrate and the working point grows with increasing component height. The working point moves further and further away from the zone of the sample that is thermally influenced by the hot plate. In order to avoid thermal fracture of the sample, the experimental setup is changed as shown in Fig. 2. Due to the oven bell, a measurement with non-contact measuring methods is now not possible. Therefore, the previous measurement results and findings are used to draw conclusions about the process inside the oven bell. A Synrad firestar f201 with $\lambda = 10.6\ \mu m$ in cw mode was used for the setup shown in Fig. 2. The heating plate was replaced by an oven bell. Inside the process chamber, a constant temperature of 500 °C was maintained during the experiments with a deviation of ± 15 K. The movement of the specimen inside the process chamber was realised with a three-axis system. Stepper motors drive the linear axes, which are controlled by a CNC card module. In this setup, the focusing unit is not moved, the constancy of the working point in the layer build-up is ensured by the z-axis of the object table. The feed direction for the experiments was perpendicular to the object table speed.

3 Results

The morphologies of the additively manufactured layers produced in the previous investigations (Fröhlich et al. 2020) confirm the results of Luo et al. (2017). By using rods with a diameter of 2 mm (borosilicate glass) and 3 mm (soda-lime glass), a higher laser power is necessary compared to Luo et al. (2017). The line energy for individual additively manufactured layers is calculated as follows (1):

$$E = \frac{P}{v_O} \quad (1)$$

The line energy refers to the laser power and the object table speed and has the unit J/mm. The ratio between the object table speed and the feed rate remains constant

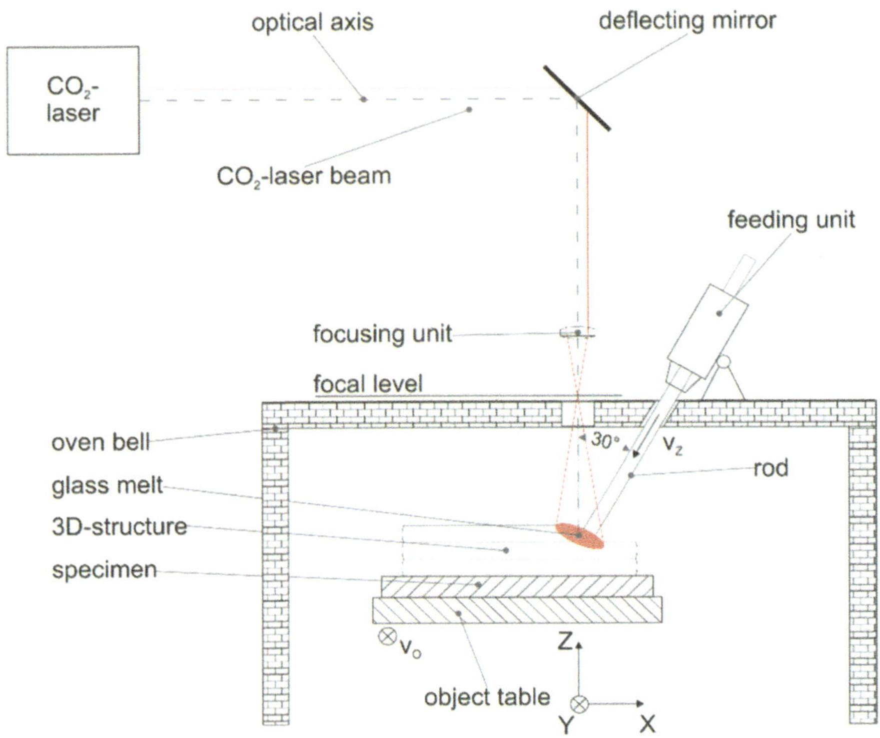

Fig. 2 Adaptation of the experimental set-up for the Direct Energy Deposition Laser process, replacing the heating plate with a furnace bell

during the experiments.

$$f = \frac{v_Z}{v_O} \tag{2}$$

Experiments conducted with the set-up (Fig. 1) had a constant ratio f = 1. The heating unit had an average temperature of 515.5 °C at its surface with a standard deviation of 7 K. In a range with a line energy of 172 J/mm to 202 J/mm, average layer widths of 5.5 mm and layer heights of 2.25 mm could be produced with borosilicate glass. It was observed that with increasing number of layers, the influence of thermal conduction and convection of the heating unit steadily decreases. The resulting thermal gradient during production increased over the height of the additively manufactured structure. This means that there is no heat accumulation in the structure. In a false colour image Fig. 3, the temperature distribution on the glass surface of the structure can be seen qualitatively. The emission coefficient for glass was $\varepsilon = 0.8$ (manufacturer's specification). The temperature curves of the ROI 1 line (Fig. 3) can be seen in Fig. 4. A striking point is the interaction area between

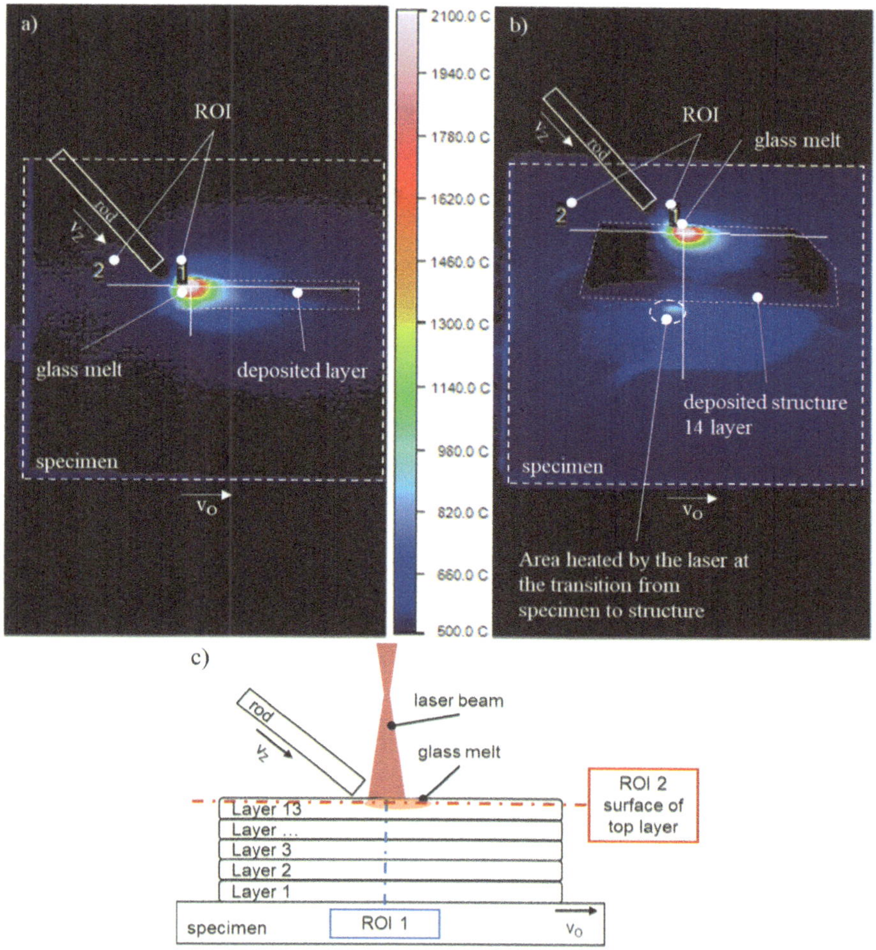

Fig. 3 False colour image of the temperature distribution on the surface of the sample. The left image **a** shows the process of the first layer. The right picture **b** shows the process during the fourteenth layer. The bottom picutre **c** shows a sketch of the 13 layer structure with ROI 1 and 2. The emission coefficient for glass was set to $\varepsilon = 0.8$ (manufacturer's specification)

the defocused laser and the specimen at the base of the structure. Each layer is exposed to a different thermal load. This leads to the remaining thermal stresses exceeding the stresses that the material can withstand during cooling or renewed abrupt heating. Figures 4 and 5 show that high temperature gradients occur in the additive manufacturing of glass.

Above the glass transition temperature T_G (Duran® $T_G = 525$ °C), it is assumed that glass has a liquid state (Jebsen-Marwedel and Brückner 2011). This explains the compensation of the prevailing temperature gradients around the centre of the glass

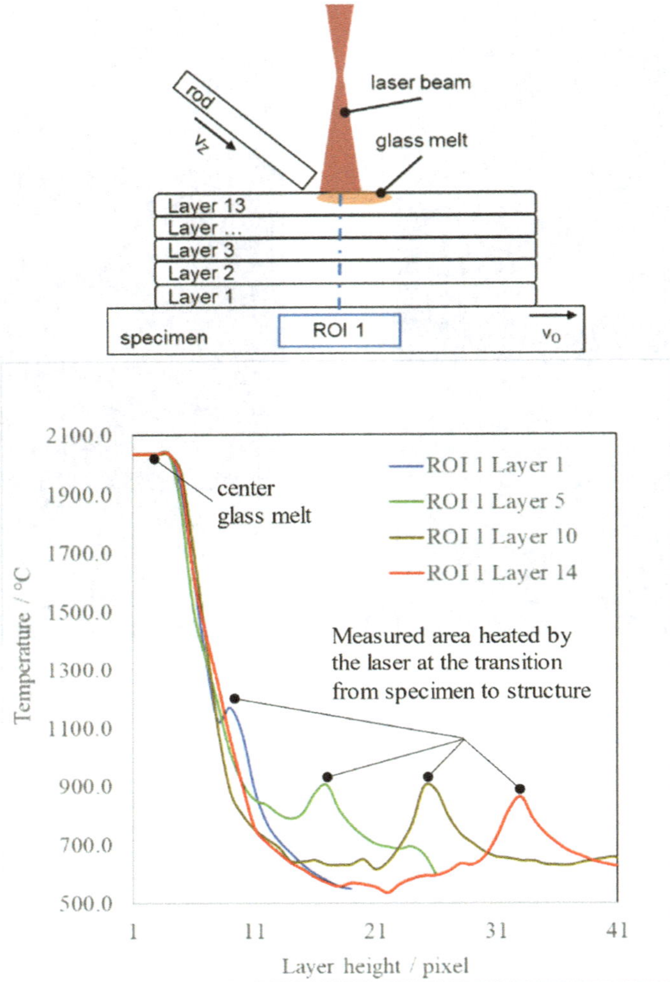

Fig. 4 The temperature curve is shown as a line profile along the ROI 1 line (Fig. 3). Measured from the centre of the glass melt over the height of the structure. No quantitative value can be assigned on the x-axis to individual pixels due to the inclination and low resolution of the thermal imaging camera

melt. In the edge regions of the structure, visible at the transition from ROI 2 layer 14 at the end of the layer (Fig. 5), the temperature gradient between structure and specimen is approximately 100 K. The further away the layer is from the heating unit, the temperature approaches room temperature and the temperature gradient becomes greater. The prevailing temperature cannot be displayed in Fig. 3 because it is outside the measuring range of the thermal imaging camera. At temperatures below the glass transition temperature T_G, it is assumed that glass takes on the behaviour of a solid

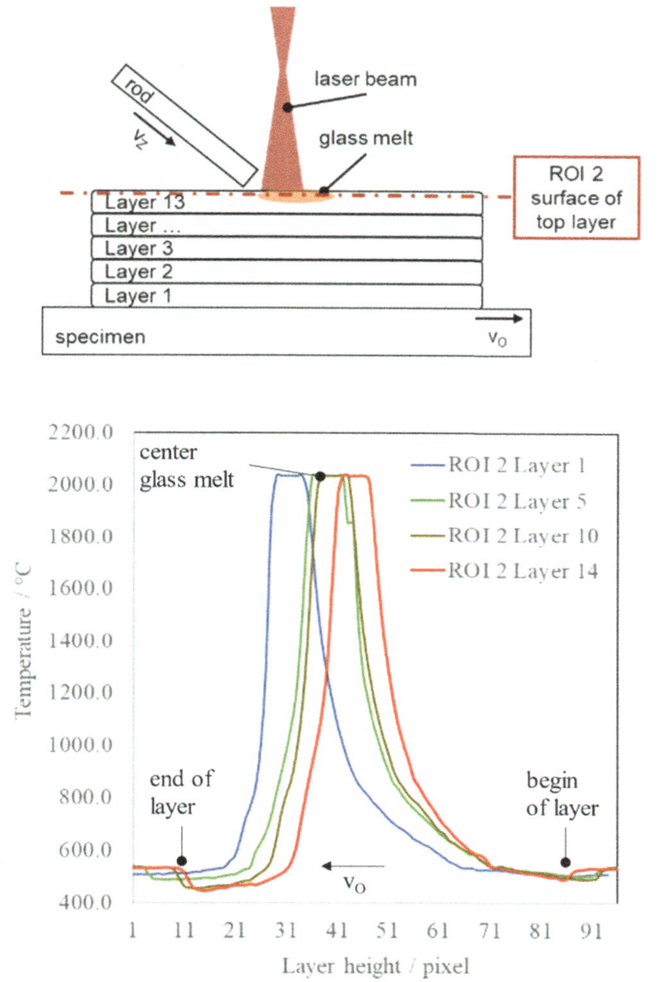

Fig. 5 The temperature curve is shown as a line profile along the ROI 2 line on the surface of the top layer. Measured from the beginning of a deposited layer to the end of deposited layer on the top of the structure. No quantitative value on the x-axis can be assigned to individual pixels due to the inclination and low resolution of the thermal imaging camera

(Jebsen-Marwedel and Brückner 2011). Thus, the tolerable temperature gradients are much lower and this has been shown during the experimental tests. In our experiments the number of possible layers ran into the limit of about 15 layers, after which the structures broke due to thermal stresses. To manufacture higher structures additively, the experimental setup was modified. The addition of a furnace bell enables a constant ambient temperature (Fig. 2). For the further experiments, the ratio f was changed to f = 1. The laser power could be reduced to P = 79 W with a variance of ± 4 watts. This corresponds to a line energy of 79 J/mm. With these values, specimens were made with both borosilicate and soda lime glass. Compared to the fabricated structures with the structure in Fig. 1, the structures with the experimental setup in Fig. 2 do not have the waviness of the applied layers similar to the FDM process. The single layers are homogeneously bonded and do not show any notches at the sides. In the transition between sample and structure, an optical separation can be seen. According to Vogel (1993) and Jebsen-Marwedel and Brückner (2011), the optical separation can be a streak in the glass, which is formed due to a viscosity difference. In the further layer structure of the single track wall (Fig. 6) such optical separations are not visible. It can be concluded from this at a homogeneous temperature field and thus a homogeneous viscosity existed. Due to the oven bell, non-contact measurement methods and the temperature prevailing there, contact measurement methods could not be used for temperature measurement. Figure 7 shows a planar application. Here, three layers

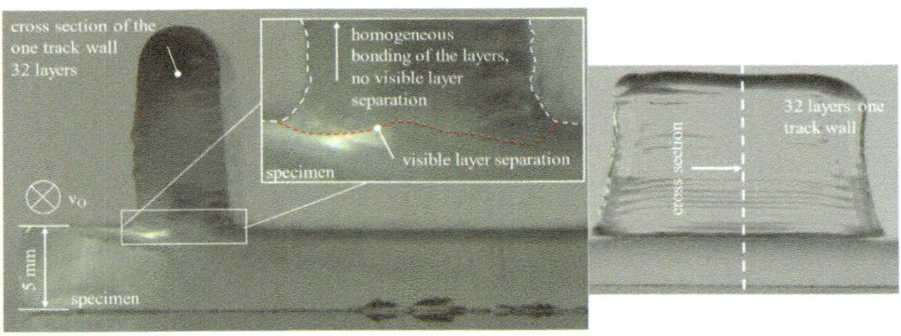

Fig. 6 Cross-section of a borosilicate glass specimen. An one-track wall with 32 layers i applied to the specimen. There is a visible layer separation at the transition between the specimen and the structure

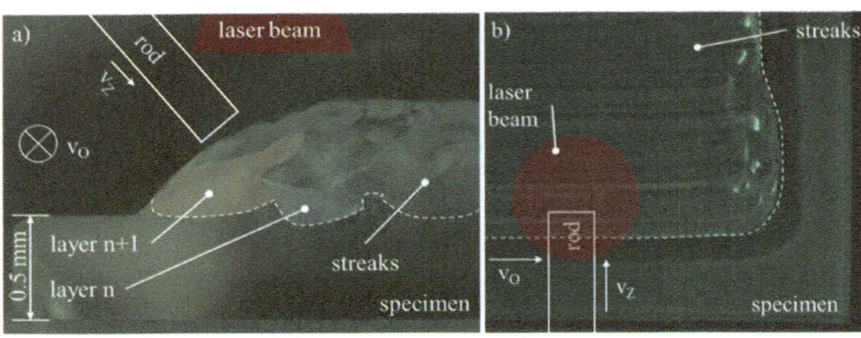

Fig. 7 Schematically, the DED-L process has been drawn in both figures. **a** Shows the cross-section of (**b**). **b** Shows the top view of the structure. The individual glass streaks can be seen, which are parallel to the direction of additiv manufacutring process. The pictures are taken under a light microscope

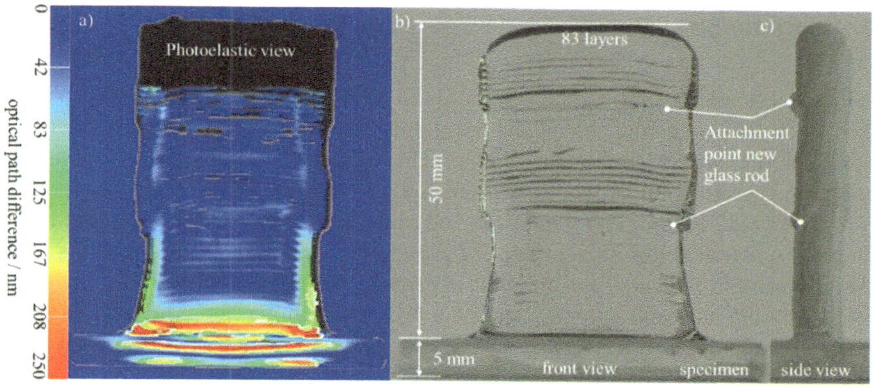

Fig. 8 Shows a component made of borosilicate glass with 83 layers and a height of 50 mm after thermal post-treatment. On the left in a, the component is examined with the photoelasticity for differences in the optical path difference in order to draw conclusions about stress concentrations. **b** Shows the front view and **c** the side view with the existing final contour

were deposited on the surface. Figure 7a shows the cross-section of Fig. 7b. Figure 7b shows the top view of the structure. The individual glass streaks can be seen, which are parallel to the direction of additive manufacturing process. Figure 7a clearly shows a mixing of the layers. The number and distinctness of the streaks indicate that different temperature conditions prevail when applying a layer to a surface compared to the construction of a single track wall. In order to prevent fracture in the structure during cooling to room temperature and to minimise thermal stresses in the component, it is thermally post-treated. After the manufacturing process, the oven was switched off and the sample cooled down with the oven. The component was then examined for existing stresses using the Strainscope S3/180 device from ilis. The optical path difference was analysed. This shows the sum of existing stresses over the thickness of the component (Fig. 8a). The wavelength of the light is accelerated or slowed down depending on the existing stress in the component. The interference of the outgoing light wave with the measured light wave indicates the optical path difference and allows conclusions to be drawn about the stresses present in the component. The black area in Fig. 8a shows the support, which cannot be penetrated by light. This was necessary to align the component perpendicular to the beam path of the photoelasticity. Figure 8b shows the front view of the manufactured component with 83 layers and a height of 50 mm. Figure 8c shows the component in side view. The attachment points of the new glass rod can be clearly seen in both pictures. In the lower third of the structure, a slightly wavy surface is only visible in the photoelastic view. A concentration of stresses is present at the transition between specimen and structure. Indicated optical path differences in the edge areas or on the wavy surface can also be due to changed reflection or transmission conditions on the curved surface.

4 Conclusion

With the extended experimental setup of the DED-L process, we were able to show that the additive manufacturing of individual large-volume glass components is possible compared to the state of the art. The use of glass rods creates a basis that coloured glass rods can also be selectively fed. Furthermore, the processing of quartz (Fröhlich et al. 2020), borosilicate and soda-lime glass has been demonstrated with the experimental setup. Other types of glass, provided they are available as rods, are conceivable with this set-up, as it is possible to adjust to the type of glass by regulating the ambient temperature in the process chamber and adjusting the laser power. The temperature measurements show that the interaction between the laser and the glass creates very high temperature gradients of up to 1500 K on the surface. In order to better distribute the resulting thermal stresses, the entire component must be kept around the glass transition temperature during production. It has been shown that streaking occurs more frequently in the glass when applied to the surface. This may be due to changing thermal boundary conditions compared to a one-track wall. To reduce the formation of streaks, further tests are planned with variation of the temperature in the process chamber, change of the ratio of the interaction area glass and laser in relation to the rod thickness and thermal simulations. The implementation of temperature measurement technology in the extended setup will help to generate more data and understanding of the prevailing temperatures in the process zone. The cooling of the sample with the switching off the furnace has shown that residual stresses are still present in the component.

Investigations with controlled cooling to reduce the stresses to a minimum are planned. A near-net-shape production could be implemented, but a post-processing is still necessary. Particular attention is being paid to the starting points of new glass rods. The waviness in the surface can be compensated by laser polishing.

Acknowledgements Part of the investigations were carried out of the project "Production and design methods for the efficient production of individual structures from silicate materials using Wire-Laser Additive Manufacturing (WLAM) -ProGlas3D". Special thanks go to the company 3D Schilling GmbH, which worked with us on the project. The Thüringer Aufbaubank EFRE (2017FE9085) funds the project. The authors are grateful to LumaSense Technologies GmbH for providing the Mikron thermal imaging camera MC320FHT.

References

Fateri, M., Gebhardt, A.: Selective laser melting of soda-lime glass powder. Appl. Ceram. **12**(1), 53–61 (2015). https://doi.org/10.1111/ijac.12338

Fateri, M., Gebhardt, A., Thuemmler, S., Thurm, L.: Experimental investigation on selective laser melting of glass. Phys. Procedia **56**, 357–364 (2014). https://doi.org/10.1016/j.phpro.2014.08.118

Fröhlich, F., Hildebrand, J., Bergmann, J.P.: Herstellung individueller Strukturen aus silikatischen Werkstoffen mittels Wire-Laser Additive Manufacturing. In: Glasbau 2020, Ernst, Wilhelm & Sohn, pp. 287–297 (2020)

Gal-Or, E., Gershoni, Y., Scotti, G., Nilsson, S.M., Saarinen, J., Jokinen, V., Strachan, C.J., Boije af Gennäs, G., Yli-Kauhaluoma, J., Kotiaho, T.: Chemical analysis using 3D printed glass microfluidics. Anal. Methods **11**, 1802–1810 (2019). https://doi.org/10.1039/C8AY01934G

Grabe, T., Lammers, M., Wang, S., Wang, X., Rettschlag, K., Sleiman, K., Barroi, A., Biermann, T., Ziebehl, A., Röttger, J., Ley. P.-P., Wolf, A., Jaeschke, P., Hermsdorf, J., Kaierle, S., Ahlers, H., Lachmayer, R.: Additive manufacturing of fused silica using coaxial laser glass deposition: experiment, simulation, and discussion. Laser 3D Manuf. **8**(11677), 45–46 (2021). https://doi.org/10.1117/12.2577205

Gräf, S., Staupendahl, G., Gerling, P., Müller, F.A.: Optical constants n and κ of various technical and optical glasses at $\lambda = 10.59\ \mu$m. J. Appl. Phys. **113**. https://doi.org/10.1063/1.4772619

Jebsen-Marwedel, H., Brückner, R.: Glastechnische Fabrikationsfehler, „Phatologische" Ausnahmezustände des Werkstoffes Glas und ihre Behebung. Eine Brücke Zwischen Wissenschaft, Technologie Und Praxis, Springer, Berlin, Heidelberg. (2011). https://doi.org/10.1007/978-3-642-16433-0

Khmyrov, R.S., Grigoriev, S.N., Okunkova, A.A., Gusarov, A.V.: On the possibility of selective laser melting of quartz glass. Phys. Procedia **56**, 345–356 (2014). https://doi.org/10.1016/j.phpro.2014.08.117

Khmyrov, R.S., Protasov, C.E., Grigoriev, S.N., Gusarov, A.V.: Crack-free selective laser melting of silicia glass: single beads and monolayers on the substrate of the same material. Int. J. Adv. Manuf. Technol. **85**, 1461–1469 (2016). https://doi.org/10.1007/s00170-015-8051-9

Kinzel, E.C., Luo, H., Pan, H.: Additive manufacturing of glass. J. Manuf. Sci. Eng. **136**(6) (2014). https://doi.org/10.1115/1.4028531

Klein, S., Simske, S., Parraman, C., Walters, P., Huson, D., Hoskins, S.: 3D Printing of Transparent Glass. HP Laboratories Technical Report (2012)

Klein, J., Stern, M., Franchin, G., Kayser, M., Inamura, C., Dave, S., Weaver, J.C., Houk, P., Colombo, P., Yang, M., Oxman, N.: Additive manufacturing of optically transparent glass. 3D Print. Additive Manuf. **2**(3) (2016). https://doi.org/10.1089/3dp.2015.0021

Klocke, F., McClung, A., Ader, C.: Direct laser sintering of borosilicate glass. Int. Solid Freeform Fabric. Symp. (2004). https://doi.org/10.26153/tsw/6986

Kotz, F., Plewa, K., Bauer, W., Schneider, N., Keller, N., Nargang, T., Helmer, D., Sachsenheimer, K., Schäfer, M., Worgull, M., Greiner, C., Richter, C., Rapp, B.E.: Liquid glass: a facile soft replication method for structuring glass. Adv. Mater. **28**(23), 4646–4650 (2016). https://doi.org/10.1002/adma.201506089

Kotz, F., Arnold, K., Bauer, W., Schild, D., Keller, N., Sachsenheimer, K., Nargang, T.M., Richter, C., Helmer, D., Rapp, B.E.: Three-dimensional printing of transparent fused silicia glass. Nature **544**, 337–339 (2017). https://doi.org/10.1038/nature22061

Lee, I., Manthriam, A., Marcus, H.L.: Selective laser sintering of alumina-zinc borosilicate glass composites using monoclinic HB02 as a binder. Int. Solid Freeform Fabric. Symp. (1995). http://hdl.handle.net/2152/68689

Liu, C., Qian, B., Ni, R., Liu, X., Qiu, J.: 3D printing of multicolor luminescent glass. Roy. Soc. Chem. Adv. **8**(55), 31564–31567 (2018). https://doi.org/10.1039/C8RA06706F

Luo, J., Gilbert, L.J., Bristow, D.A., Landers, R.G., Goldstein, J.T., Urbas, A.M., Kinzel, E.C.: Additive manufacturing of glass for optical applications. In: Proceedings 9378, Laser 3D Manufacturing III (2016). https://doi.org/10.1117/12.2218137

Luo, J., Gilbert, L.J., Qu, C., Landers, R.G., Bristow, D.A., Kinzel, E.C.: Additive manufacturing of transparent soda-lime glass using a filament-fed process. J. Manuf. Sci. Eng. **136**(6) (2017). https://doi.org/10.1115/1.4035182

Luo, J., Hostetler, J.M., Gilbert, L.J., Goldstein, J.T., Urbas, A.M., Bristow, D.A., Landers, R.G., Kinzel, E.C.: Additive manufacturing of transparent fused quartz. Opt. Eng. **57**(4) (2018). https://doi.org/10.1117/1.OE.57.4.041408

Nguyen, D.T., Meyers, C., Yee, T.D., Dudukovic, N.A., Destina, J.F., Zhu, C., Duoss, E.B., Baumann, T.F., Suratwala, T., Smay, J.E., Dylla-Spears, R.: 3D-printed transparent glass. Adv. Mater. **29**(26) (2017). https://doi.org/10.1002/adma.201701181

Schaeffer, H.A., Langfeld, R.: Werkstoff Glas. Alter Werkstoff Mit Großer Zukunft, Technik Im Fokus. (2020). https://doi.org/10.1007/978-3-662-60260-7

Seel, M., Akerboom, R., Knaack, U., Oechsner, M., Hof, P., Schneider, J.: Additive manufacturing of glass components—exploring the potential of glass connections by fused deposition modeling, challenging glass 6—conference on architectural and structural applications of glass. https://doi.org/10.7480/cgc.6.2161

Vogel, W.: Glasfehler, Springer, Berlin. Heidelberg (1993). https://doi.org/10.1007/978-3-642-58048-2

Witzendorff, P., Pohl, L., Suttmann, O., Heinrich, P., Heinrich, A., Zander, J., Bragard, H., Kaierle, S.: Additive manufacturing of glass: CO_2-Laser glass deposition printing. Procedia CIRP **74**, 272–275 (2018). https://doi.org/10.1016/j.procir.2018.08.109

Adhesive Bonding

Characterization of Layer Properties for Components Made of Lignin Based Filaments Manufactured by Material Extrusion

M. Fiedler, F. Fischer, and K. Droeder

Abstract The reduction of petroleum-based components is an important issue in the polymer-industry. The long-term-effects of using these materials and their disposal procedures on the environment are critical. The main focus is increasing due to the use of renewable, wood- and bio-based raw materials in polymeric matrices. In this study, lignin-sulfonate (LGS) is used, a cheap feedstock from biomass, to develop sustainable polymers for consumption in additive manufacturing (AM). Today, there is very limited utilization of wood-based materials substituted for petroleum-based polymers for the fused deposition modeling (FDM). As petroleum-based polymers are used often, the purpose is to find a sustainable composite for FDM. Therefore, the polymer acrylonitrile–butadiene–styrene (ABS) is combined with lignin as a next generation bio-based composite. To determine the thermal properties, the mentioned materials ABS and LGS are first examined using thermal gravimetric analysis (TGA) and differential scanning calorimetry (DSC). In order to determine whether the addition of lignin has a positive effect on layer adhesion in the FDM process, the results are compared with a conventional ABS Filament by using defined parameters. The main objective of adding bio-based materials in polymeric matrices for creating a printable green filament with sufficient adhesion properties can be made. The results suggest that bio-based filaments can be made as a more sustainable and low-cost material.

Keywords Additive manufacturing · Sustainable materials · Lignin · Bio based · Wood · Fused deposition modeling

M. Fiedler (✉) · F. Fischer
Volkswagen Aktiengesellschaft, Berliner Ring 2, 38436 Wolfsburg, Germany
e-mail: maximilian.fiedler1@volkswagen.de

F. Fischer
e-mail: fabian.fischer2@volkswagen.de

K. Droeder
Institute of Machine Tools and Production Technology, Technical University of Braunschweig, Langer Kamp 19b, 38106 Braunschweig, Germany
e-mail: k.droeder@tu-braunschweig.de

L. F. M. da Silva et al. (eds.), *2nd International Conference on Advanced Joining Processes (AJP 2021)*, Proceedings in Engineering Mechanics,
https://doi.org/10.1007/978-3-030-95463-5_11

1 Introduction

The use of sustainable components is a major topic in automotive applications. The renewable material wood, being used in combination with various materials such as technical polymers, offers a high potential for producing innovative components and alternative automotive materials with an improved CO_2 footprint. This research and development are essential for aiming the global climate goals. As a first step into emission-free mobility, the innovation of sustainable components based on circular economy techniques in the automotive industry will be explored. Next to new solutions for vehicle concepts and energy sources, the topics weight reductions and sustainable production are also essential for a future-oriented and ecological mobility (Mainka et al. 2015).

Technical requirements for future components in the automotive industry focusing on lightweight construction, resource- and cost efficiency. The use of renewable raw materials, such as wood and bio-based polymers will increase. Several studies mentioned the use of lignin, as a waste product of the pulp and paper industry with a high production volume, for industrial applications (Asada et al. 2005; Cazacu et al. 2004; Thielemans and Wool 2005). The sustainable material wood, as the most common renewable material in our environment, offers a high potential for applications in the automotive industry with regard to the specific mechanical properties. Used in combination with technical polymers the hybrid structure can be further optimized and adjusted to component-specific requirements (Song et al. 2018).

Lignin, as waste product from the pulp and paper industry and next to cellulose the second most abundant plant based renewable material in wood structures, has been a research topic for more than 50 years (Freudenberg and Neish 1968). It's heterogeneous structure and non-uniform quality concludes in different chemical and physical modification steps to convert it in an effective product. As a result, the associated processes become more energy and also cost intensive and the successful commercialization failed (Kadla et al. 2002; De Wild et al. 2009; Azadi et al. 2013; Ragauskas et al. 2014; Steward 2008). During the last decades the integration of lignin within polymeric matrices enabled the preparation of lignin-based composites for industrial applications. But these applications focusing on conventional manufacturing methods such as injection molding or casting (Thakur 2014; Rozman et al. 2000; Toriz et al. 2002; Tran et al. 2016; Saito et al. 2012).

The Additive Manufacturing (AM) technologies have spread rapidly in the last decade. The most common used technology in AM is the Fused Deposition Modeling (FDM). FDM uses thermoplastic filaments melted at high temperature and solidified when cooling. In most use cases the objects are built from the bottom to the top (Go and Hart 2016; Kalsoom et al. 2016). All processes are digitized and computer-controlled, no additional costs are required for design complex geometries and structures. Less waste and chemicals are produced in comparison to traditional manufacturing processes. Therefor FDM is a promising and sustainable technology with high energy efficiency and low environmental effects (Wimmer et al. 2015; Campbell et al. 2011). Actually, there is very low utilization of bio-based materials substituted for petroleum-based polymers (Bhatia and Ramadurian 2017). In addition to the mentioned low environmental effects by using these new techniques, there is now a growing interest in bio-based, bio-compatible and bio-added material for 3D-printing. Unlike regular thermoplastics, wood-based materials are challenging to melt processes (Nguyen et al. 2018; Wimmer et al. 2015; Bhatia and Ramadurian 2017; Gardan and Roucoules 2014; Henke and Treml 2013).

Due to its mechanical performance and good FDM compatibility the thermoplastic polymer Acrylonitrile butadiene styrene (ABS) is one of the most common used 3D-printing filaments (Bhatia and Ramadurian 2017; Nguyen et al. 2018).

In this study Lignin-sulfonate (LGS) and petrol-based ABS was composed to develop a sustainable filament for the FDM process. Due to their different chemical-, thermal- and mechanical properties the objective was to create a suitable process for combining those materials. At the end the developed material had comparable mechanical performance to common petroleum-based materials and excellent 3D-printability. As a limitation, it has to be mentioned that the olfactory perception of the manufactured objects has to be classified as negative. The sulphonic acid group of the LGS, which is shown in Fig. 1, emits the characteristic odor of sulfur.

At first, the manufacturing process of lignin-based filament is investigated. To determine the thermal properties, the mentioned materials ABS and LGS are first examined using thermal gravimetric analysis (TGA) and differential scanning calorimetry (DSC). Based on the thermal properties, both materials are mixed in different weight percent, with a lignin content up to 40 weight percent (WT%). Important tensile tests according to DIN EN ISO 527-2 were printed and the microscopic examination of the fractured surface are made.

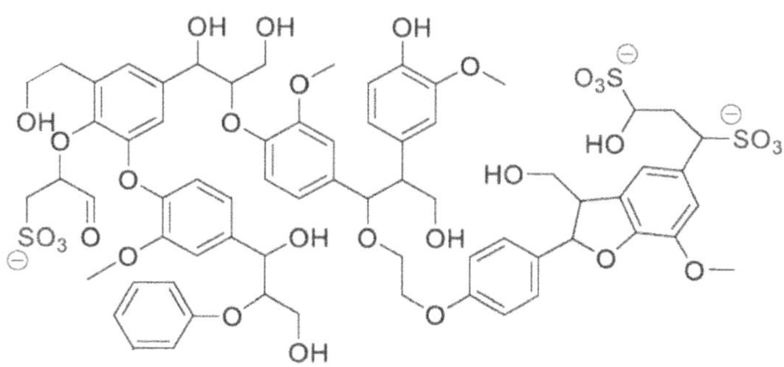

Fig. 1 Chemical structure of lignin-sulfonate

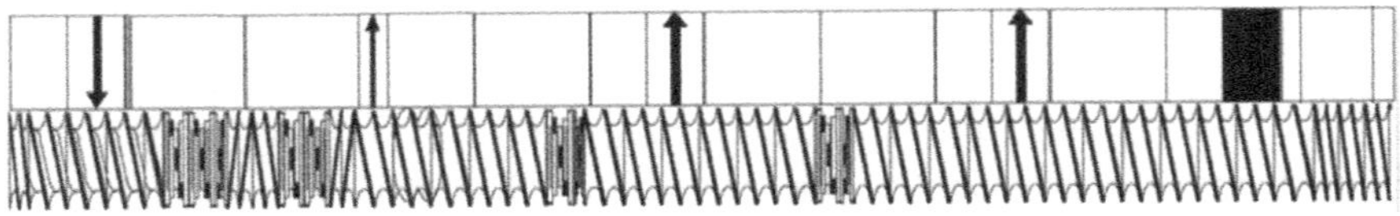

Fig. 2 Configuration of the extruder-screw

2 Experimental

2.1 Materials

The used ABS is a conventional polymeric filament for the FDM process. The used filaments were commercialized by Ultimaker Holding B.V. and purchased from a reseller for 3D-Printing applications. The used lignin is magnesium lignosulphonate gained from the magnesium bisulphite process. The lignin was sourced by the paper production at Sappi Biotech GmbH in Germany.

2.2 Composite Processing

The filament has been shredded into pellets and both materials were mixed at 210 °C and 300 rpm by using a Coperion ZSK18 ML extruder (Fig. 2). LGS powder is added to the process behind the solid-feed section and was added after the ABS. The mixing time was at least 20 min and four different composites were made out of lignin. In each process the amount of Lignin in the filament was increased of about 10 wt%. It was started at 10 wt% Lignin at ends with 40 wt% Lignin.

The manufactured blends were stored at room temperature in a closed tank. After a mandatory drying process, the blends were manufactured by a 3Devo Precision 350 extruder to a proper 3D-Printing filament. Due to the four adjustable heating-zones the material was extruded between 205 and 220 °C. The screw works at about 7 rpm and the filament diameter was set to 2.85 mm ± 0.1 mm.

2.3 Thermal Analysis

The tests are made at normal climate, the preconditioning was done according to DIN EN ISO 291 at 23 °C and relative humidity of 50% for at least 16 h. Thermal characteristics of all samples were quantified by a thermo-gravimetric analyzer (TG 209 F1 Libra Netzsch-Gerätebau GmbH) according to DIN EN ISO 11358 with sample weight of 90 mg. To remove moisture, the sample was loaded and kept isothermal at 100 °C for 20 min. From 30 to 600 °C the measurements were performed in nitrogen atmosphere, then ramped to 900 °C by using oxygen atmosphere and a ramp rate of 20 K/min.

By using a differential scanning calorimeter (DSC 204 F1 Phoenix ASC Netzsch-Gerätebau GmbH) the glass transition temperature (Tg) and heat capacity change (Cp) were investigated. The measurements were performed according to DIN EN ISO 11357 in nitrogen atmosphere. After removing the thermal history with the first cycle, the second cycles were almost identical. To measure the Tg and Cp of these samples the second cycles were used.

2.4 Printing Process

The filaments are printed to specimens in accordance to DIN EN ISO 527 Type 1A using the FDM process. They are printed in the xy-plane and the z-direction, also shown in Fig. 3, to compare the mechanical properties of the specimens in different printing directions.

In general, low printing speeds and higher temperatures produce better results (Johansson 2016). Therefore, an Ultimaker equipped with a 0.4 mm diameter nozzle was used to print the filaments and the chosen geometries were sliced by an Ultimaker specific Cura-Software. To avoid possible warping effects when printing with ABS the fan-speed was set to 0% and the Ultimaker's building space is completely capsuled for better layer adhesion. Due to the results from the thermal analysis, we used a specific set of printing parameters in this study. The printing temperature was set to 230 °C, bed temperature was set to 110 °C in accordance to the measured glass transition temperature and the sprinting speed was about 15 mm/s to avoid damages by printing in z-direction due to leveraging effects.

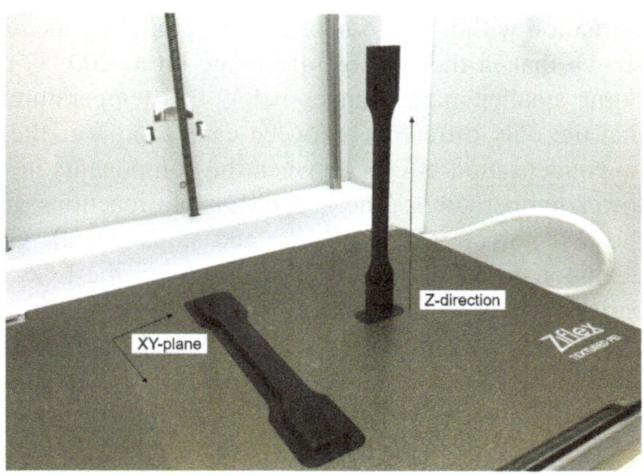

Fig. 3 Printing strategies for specimens

2.5 *Mechanical Measurements*

The tensile tests are also made at normal climate, the preconditioning was done according to DIN EN ISO 291 at 23 °C and relative humidity of 50% for at least 16 h. The tensile mechanical properties of the studied materials were received with a uniaxial fixture using the Z250 Xtens ZwickRoell. Test conditions and parameters were chosen according to DIN EN ISO 527–2. The measured data were analyzed by using the TestXpert software provided by ZwickRoell.

2.6 *Optical Analysis*

In order to gain a deeper level of knowledge for the characterization of the material, microscopic analyses were performed. The filaments used to produce the test specimens, the fractured edges and the surfaces of the tensile specimens were evaluated. The filaments were inspected by an Axio Imager 2 microscope from the Carl Zeiss Microscopy GmbH.

3 Results and Discussion

3.1 *Thermal Properties*

In Fig. 4 the thermograms from the TGA were shown to evaluate the thermal stability of ABS in comparison with different lignin contents. The residual mass at 900 °C is shown for each filament. The material properties strongly depend on the structure formation within the composite. By comparing the different adhesive systems, it is shown that in the temperature range up to 200 °C the composites remain stable. Some smaller mass changes below this temperature takes place. Due to the fact, that the only difference between each sample is the amount of LGS, there might be condensation effects between the components in which water evaporates at the lower temperatures. The second visible reactions are shown at about 270 °C as the first polymeric degradation and stops at 420 °C. That means, that the curves deviation from the pure ABS depends on the LGS content and the relocation of early thermal degradation takes place by increasing the amount of LGS at lower temperatures. From about 420 to 600 °C the curves split up into four different mass change stages, which cause on an increase of the lignin content. The thermal stability of the compound system increases as the percentage of lignin increases. Above temperatures of 600 °C, a plateau is reached, that means a constant mass sets up, at that point the atmospheric gas was changed from nitrogen to oxygen in order to burn off the remaining components. The remaining ash, is an an-organic residual mass. For larger lignin contents, the residual mass correlates with the LGS content

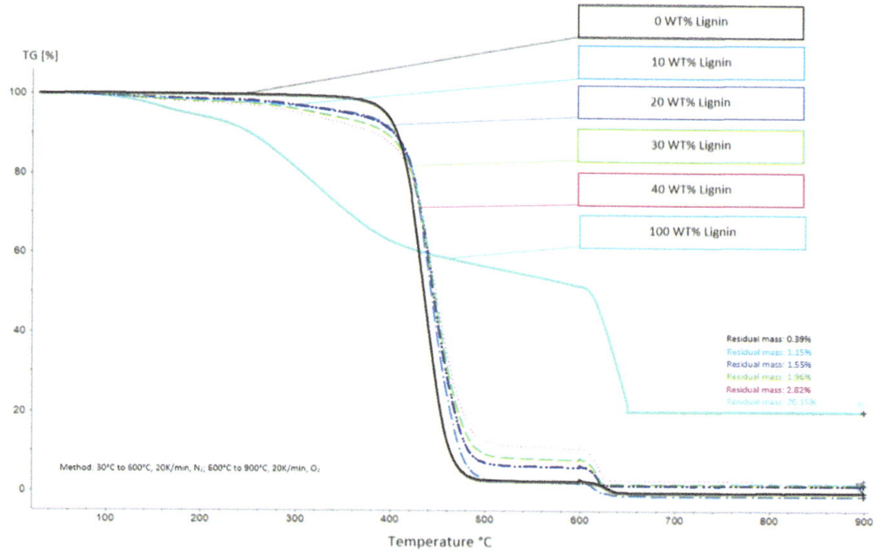

Fig. 4 TGA curve of the ABS filaments with different lignin contents

in the matrices. While the reference sample has a residual mass of about 0.0%, the sample with 40 wt% lignin has a residual mass of about 3%. The sensitivity to high temperatures decreases, due to the shown high temperature durability of the polymer structure. The thermo-gravimetric results show that LGS can only slightly improve the thermal stability of the ABS at high temperatures.

The flow behavior an the thermal transitiom of these composites are critical factors directly affecting their usability in 3D-printing applications (Nguyen et al. 2018). Figure 6 represent the thermal characteristics for the manufactured blends from 10 wt% lignin up to 40 wt% lignin. The step change at about 110 °C indicates the glass transition temperature (Tg). In this study two independent runs for each sample were made. For the characterization of Tg we only take a look at the second cycle. The first cycle was used to remove the thermal history. With higher amounts of lignin, the glass transition temperature moves slightly to the left side of the diagram. Higher lignin contents cause slightly lower Tg. As all temperatures vary by a maximum of one degree, the deviation can also be caused by measurement inaccuracies. With the use of DSC, it was possible to conclude that high mixtures of ABS and lignin are possible and that the bonding of the two materials takes place in partial amounts. The peak in the evaluation curve indicates lignin, since ABS as a semi-crystalline polymer does not have a melting peak, such a peak is characteristic for the first heating curve of lignin. The peak is the residual moisture of the hygroscopic lignin that escapes from the material by evaporation. At this moment a melting peak is simulated. The amount of required energy in relation to the mass of lignin is greater in the reference than in the composites. This can be explained by the fact that lignin in the composites has already been seen several heating cycles and therefore contains less residual moisture per wt%. Lignin behaves more duroplastic than thermoplastic, no glass transition or melting peak of the material can be identified in the DSC curve. The glass transition of the ABS does not change despite the addition of lignin.

During the utilization of lignin in the compound materials, it must be ensured that the utilization temperatures are aligned according to the thermally more unstable

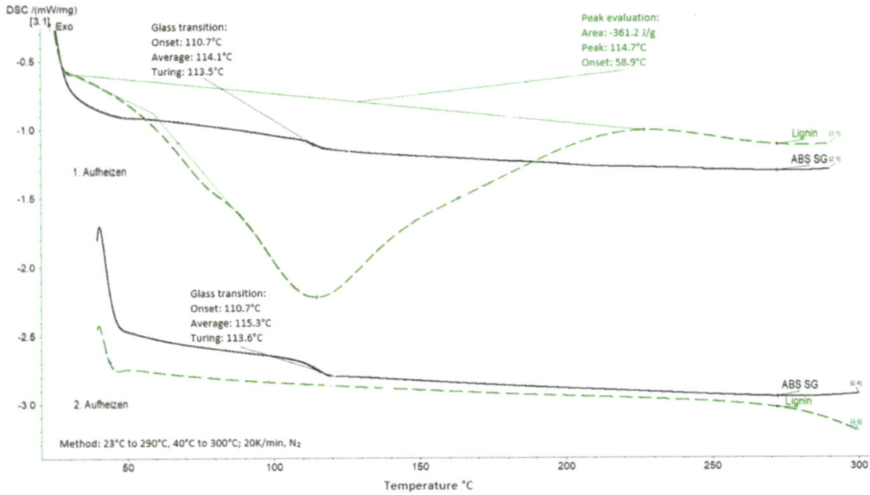

Fig. 5 Measured heat flow in DSC for ABS and LGS

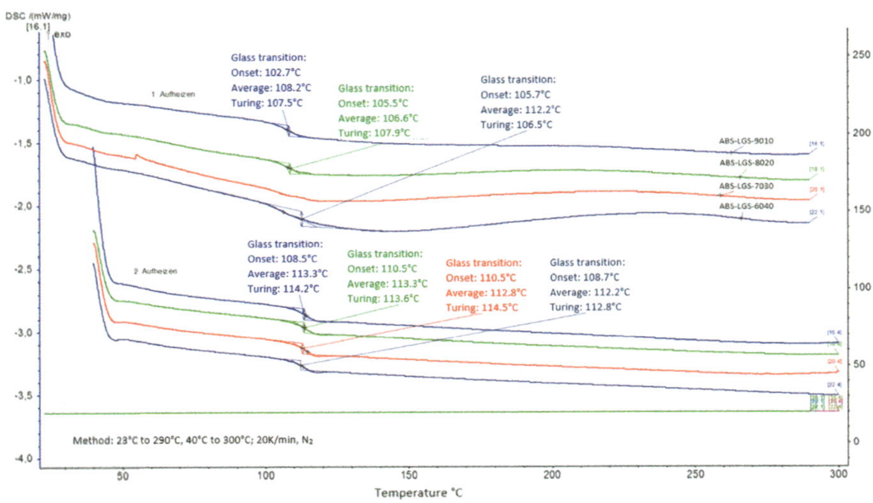

Fig. 6 Measured heat flow in DSC for different wt% LGS

component, the lignosulphonate. The stability of the material mixed with ABS, is lower than the pure ABS without further additives. The lignin did not deteriorate during processing into a composite at temperatures up to 250 °C for a shorter period of time. The decomposition of the lignin starts at a temperature of 180 °C, the decomposition of the ABS starts at 350 °C. By fabricating this composite at higher temperatures, the properties of the pure lignosulphonate will be reduced due to the absence of the intact component. Due to the thermal stability of the aromatic structures in the lignin, the degradation of the lignin starts early but nevertheless a very high residual mass remains.

3.2 *Mechanical Performance*

The tensile specimens were printed and tested in accordance to DIN EN ISO 527-2. Therefor the manufactured blends, as mentioned above, were manufactured by a filament extruder to a proper 3D-Printing filament. Due to the four independent adjustable heating-zones we decided to extrude the material between 205 and 220 °C (Fig. 7). The screw turns with 7 rpm and the filament diameter was set to 2.85 mm ± 0.1 mm (Fig. 8). Filaments can only be printed without errors if the diameter deviations are smaller than ±0.10 mm. Diameter deviations greater than ±0.10 mm, results

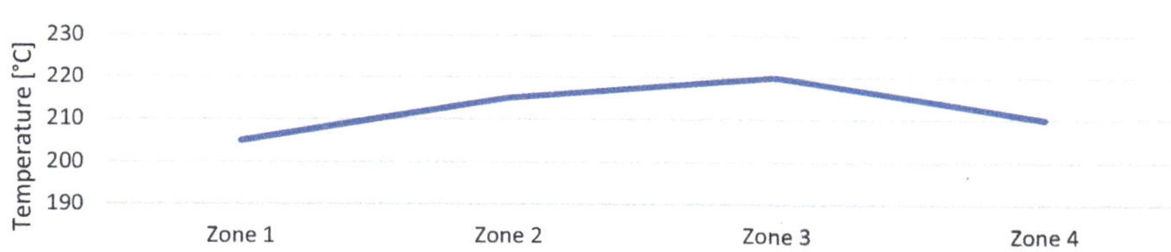

Fig. 7 Temperature profile for material extrusion

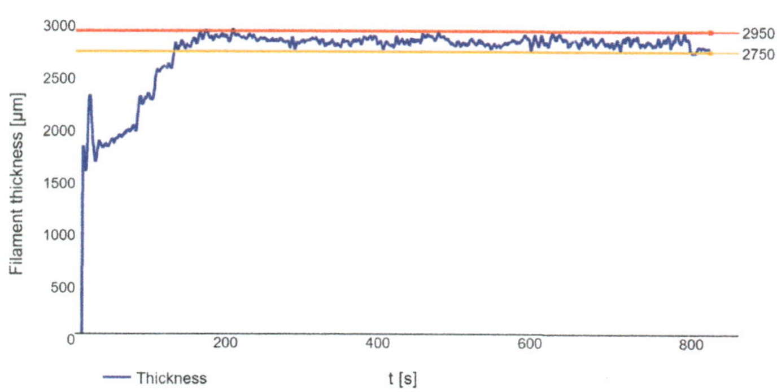

Fig. 8 Filament diameter

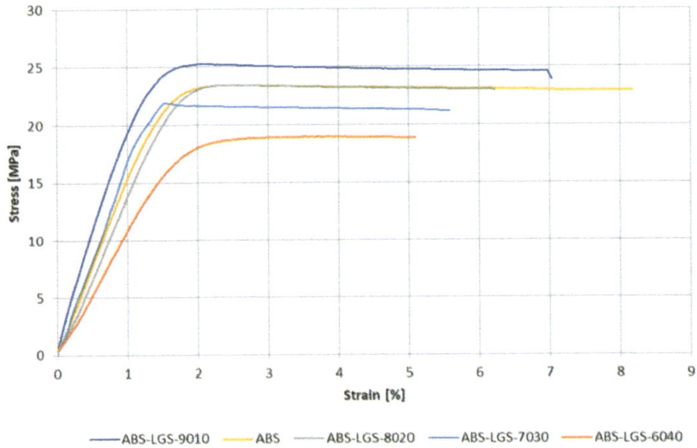

Fig. 9 Tensile test data xy-direction

in clogging printer nozzles or over-extrusion during printing. Too small diameters leads in lower extrusion rates and defects due to missing material in the component.

By modifying the ABS-based FDM filament with lignosulphonate we are able to produce materials with advanced mechanical performance (Fig. 9). For lower lignin contents (<20 wt%) the good elongation characteristics of ABS, combined with the aromatic structures of lignin causes in reinforcement effects with weaker ductility. Adding high amounts of lignin in the ABS matrix causes in lower ductility and brittle structures with less performance than the pure ABS (Fig. 9). The possible tensile strain of ABS-LGS-6040 is much smaller, than the pure ABS filament. For ABS-LGN-9010 the Tensile Young's modulus raises from 1.42 ± 0.24 GPa for pure ABS up to 2.19 ± 0.06 GPa. Due to lignin's very low molecular weight, adding high amounts (>20 wt%) in the ABS matrix causes in brittleness. Analogous to the mentioned strain-rates, the Tensile Young's modulus decreases from 1.42 ± 0.24 GPa for the pure ABS down to 0.85 ± 0.23 GPa (Fig. 10). The specimens printed in the z-direction have much lower stress- and strain-rates than the specimens printed in the xy-plane. The strain rates from the specimens printed in z-direction are six times lower compared to specimens printed in xy- plane (Figs. 10, 11, and 12). This indicates that the layer adhesion also decreases with increasing lignin content, since the individual layers separate more easily from each other due to delamination.

3.3 Optical Analysis

In the microscopic images, the lignin is easily identifiable as a component of the composite due to some agglomerates (bright dots in the filament cross-section). It can be concluded that ABS and lignin do not react with each other. In Fig. 13

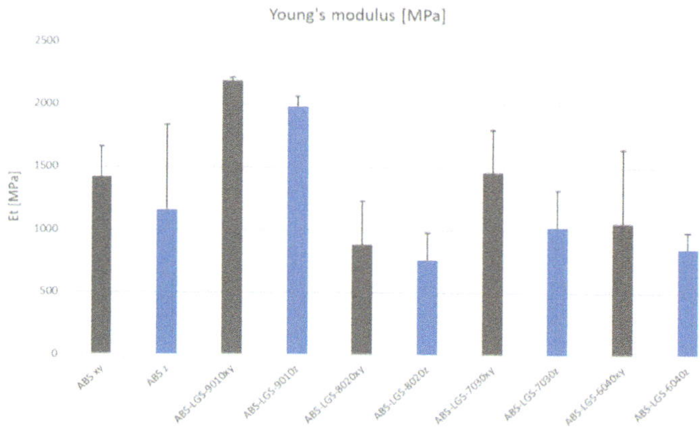

Fig. 10 Tensile Young's modulus

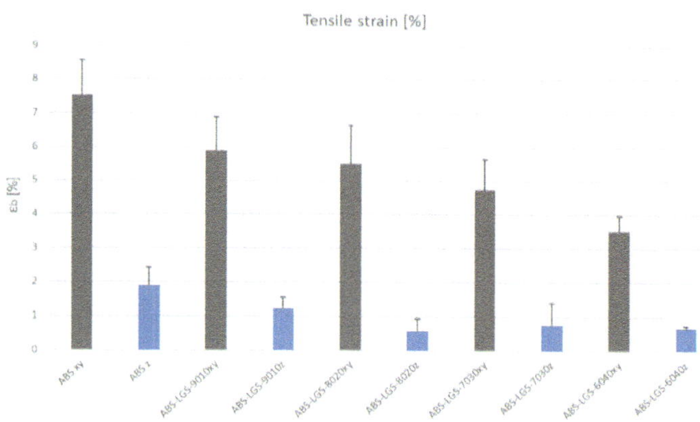

Fig. 11 Strain rates in comparison to xy- and z-direction

ABS-LGS-9010 and ABS-LGS-6040 are shown. In the filament with 40 wt% LGS, more larger agglomerates (light brown dots) are visible, compared to the filament with 10 wt% LGS. In addition, the material is more porous, which can be seen by the black dots. It can also be seen that the filament becomes more porous with a larger mass fraction of lignin. This can be seen from the lighter discoloration of the cross-section and the fine pores. Lignin is therefore only enclosed by the matrix. However, intermolecular interactions can occur between the materials. The so-called π-π-stacking results in an interaction between the aromatic structures of ABS and lignin. Due to the delocalized electrons in the aromatic compounds, mesomerism effects occur, leading to high binding forces (Kubik 2005).

Material strengthening effects can be demonstrated for the specimens produced with the addition of a maximum of 20 wt% LGS in the xy-plane. The specimens

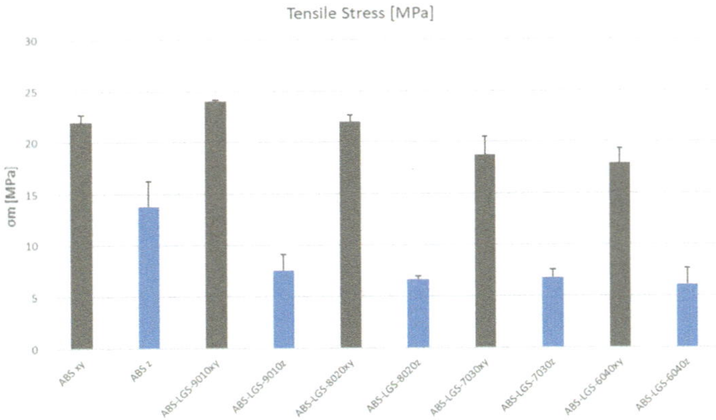

Fig. 12 Stress rates in comparison to xy- and z-direction

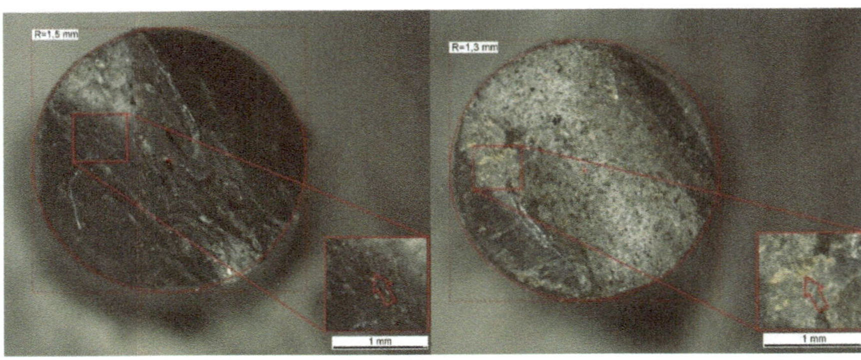

Fig. 13 Comparison of ABS-LGS-9010 and ABS-LGS-6040

prepared in the z-direction experience rapidly worsening mechanical properties with the addition of LGS. Premature material failure occurs, caused of weak layer adhesion. This can also be seen in the microscopic analysis of the fracture surfaces (Fig. 14). The fractures of the z-specimens are located between the layers. Though up to three layers are usually affected by the fracture, the layers are clearly identifiable and not affected in a continuous transition.

4 Conclusion

In this work, the mechanical and thermal properties of lignin-based compounds are characterized for FDM printing in comparison to petroleum-based filaments. The composition of lignin filaments has an effect on the properties of the composite

Fig. 14 Fractured surfaces

material and change the mechanical characteristics. It can be shown, that lignin as a bulk material in petroleum-based matrices can increase strength. The microscopic test of the produced filaments shows inhomogeneities by high level lignin contents due to agglomeration effects. Higher the lignin contents in the filament, causes in more frequently defects and blockages in the printer nozzle. Furthermore, lignin appears to be highly abrasive. The printer nozzle had to be replaced after several prints of the filament in order to be able to print further components with the material. The strong directional dependence of the printed structures also must be taken into account. This can be solved by constructive modifications to the geometry. Altogether, it shows that the bio-based additive lignin can be added to conventional ABS filaments up to high lignin amount of 40 wt% and this improves particularly the mechanical properties and the use of hybrid bio-based materials.

Funding The author(s) received no financial support for the research, authorship, and/or publication of this article.

Declaration of Conflicting Interests The author(s) declared no potential conflicts of interest with respect to the research, authorship, and/or publication of this article.

ORCID ID https://orcid.org/0000-0002-7029-5348

References

Asada, C., Nakumura, Y., Kobayashi, F.: Waste reduction system for production of useful materials from unutilized bamboo using steam explosion followed by various conversion methods. Biochem. Eng. J. **23**(2), 131–137 (2005)

Azadi, P., Inderwildi, O., Farnood, R., King, D.: Liquid fuels, hydrogen and chemicals from lignin: a critical review. Renew. Sustain. Energy Rev. 506–523 (2013)
Bhatia, S., Ramadurian, K.: 3D Printing and Bio-Based Materials in Global Health. Springer Link (2017)
Campbell, T., Williams, C., Ivanova, O., Garrett, B.: Could 3D Printing Change the World? Technologies, Potential and Implications of Additive Manufacturing. Atlantic Council (2011)
Cazacu, G., Mihaies, M., Pascu, M., et al.: Polyolefin/ligninsulfonate blends. Ind. Crops Prod. **20**(204), 205–219 (2004)
De Wild, P., Van de Laan, R., Kloekhorst, A., Heeres, E.: Lignin valorisation for chemicals and fuels for pyrolysis and hydrodeoxygenation. Environ. Prog. Sustain. Energy 461–469 (2009)
Freudenberg, K., Neish, A.: Constitution and biosynthersis of lignin. Springer Verlag (1968)
Gardan, J., Roucoules, L.: 3D printing device for numerical control machine and wood deposition. J. Eng. Res. Appl. 123–131 (2014)
Go, J., Hart, A.: A framework for teaching the fundamentals of additive manufacturing and enabling rapid innovation. Addit. Manuf. 76–87 (2016)
Henke, K., Treml, S.: Wood based bulk material in 3D printing processes for applications in construction. Eur. J. Wood Wood Prod. **71**, 139–141 (2013)
Johansson, F.: Optimizing Fused Filament Fabrication 3D Printing for Durability. Tensile Properties & Layer Bonding. Blenkinge Institute of Technology (2016)
Kadla, J., Kubo, S., Venditti, R., Gilbert, R., Compere, A., Griffith, W.: Lignin based carbon fibres for composite fibre applications. Carbon 2913–2920 (2002)
Kalsoom, U., Nesterenko, P., Paull, B.: Recent developments in 3D printable composite materials. RSC Adv. (2016)
Kubik, S.: *Lehre - Supramolekulare Chemie - Intermolekulare Wechselwirkungen*. https://www.chemie.uni-kl.de/fachrichtungen/oc/kubik/index.php?lan=de&lev1=0tea&lev2=oc9&lev3=all (2005)
Mainka, H., Täger, O., Körner, E., et al.: Lignin—an alternative precursor for sustainable an cost-effective automotive carbon fibre. J. Mater. Res. Technol. 283–296 (2015). Elsevier Editora Ltda
Nguyen, N., Bowland, C., Naskar, A.: A general method to improve 3D-printability and inter-layer adhesion in lignin-based composites. In: Applied Materials Today 12, pp. 138–152. Elsevier, Oak Ridge (2018)
Ragauskas, A., Beckham, G., Biddy, M., Chandra, R., et al.: Lignin valorization: improving lignin processing in the biorefinery. Science 709–720 (2014)
Rozman, H., Tan, K., Kumar, R., Abubakar, A., et al.: The effect of lignin as a compatibilizer on the physical propersties of coconut fiber-polypropylene composites. Eur. Polym. J. 1483–1494 (2000)
Saito, T., Brown, R., Hunt, M., Pickel, J., Messmann, J., et al.: (2012). Turning renewable resources into value-added polymer: development of lignin-based thermoplastic. Green Chem. 3295–3303 (2000)
Song, J., Chen, C., Zhu, S.: Processing bulk natural wood into a high-performance structural material. Nature **554**, 224–228 (2018)
Steward, D.: Lignin as a base material for material applications: chemistry, application and economics. Indus. Crops Prod. 202–207 (2008)
Thakur, V., Thakur, M., Raghavan, P., Kessler, M.: Progress in green polymer composites from lignin for multifunctional applications: a review. Chem. Eng. (2014)
Thielemans, W., Wool, R.: Lignin esters for use in unsaturated thermosets: lignin modification and solubility modeling. Biomacromolecules (2005)
Toriz, G., Denes, F., Young, R.: Lignin-polypropylene composites. Part 1: composites from unmodified lignin and polypropylene. Polym. Compos. (2002)
Tran, C., Chen, J., Keum, J., et al.: A new class of renewable thermoplastics with extraordinary performance from nanostructured lignin-elastomers. Adv. Funct. Mater. (2016)
Wimmer, R., Steyer, B., Woess, J., et al.: 3D Printing and Wood. Pro Ligno (2015)

Practical Implementation and Validation of a Novel Process for Manufacturing Milling Tools Using Adhesive Bonding

D. S. Correia, E. A. S. Marques, R. J. C. Carbas, P. J. C. das Neves, and L. F. M. da Silva

Abstract Although adhesive bonding is widely used in many industries, its application in the wood milling tool sector is still limited. Nonetheless, it has the potential to represent a valid alternative to the brazing process. To this end, adhesive joints must fulfil three main requirements: use an induction curing adhesive to increase production flexibility; guarantee enough electrical conductivity in the joint for the electrical discharge machining (EDM) sharpening process; and finally ensuring that all safety standards are respected. To validate the practical applicability of this joining process, an induction-based cure cycle was studied by characterising the adhesive polymerisation recurring to differential scanning calorimetry (DSC) tests. The electrical conductivity was validated in real milling tools by applying an EDM sharpening process. As for the safety requirements, the test conditions were set by the safety standard EN 847-1, where a rotational speed test at an overspeed condition was employed, validating its performance if no joint failure occurs. These results allowed to determine the applicability of the bonded tool, enabling its performance to be compared with that of the original brazed configuration. Although important

The original version of this chapter was revised: Incorrect author names have been corrected online.
The correction to this chapter is available at
https://doi.org/10.1007/978-3-030-95463-5_13

D. S. Correia (✉) · E. A. S. Marques · R. J. C. Carbas
Institute of Science and Innovation in Mechanical and Industrial Engineering (INEGI), Porto, Portugal
e-mail: dcorreia@inegi.up.pt

E. A. S. Marques
e-mail: emarques@fe.up.pt

R. J. C. Carbas
e-mail: rcarbas@fe.up.pt

P. . J. C. das Neves
FREZITE – Woodworking Tools, SA, Trofa, Portugal
e-mail: paulo.neves@frezite.pt

L. F. M. da Silva
Department of Mechanical Engineering, Faculty of Engineering (FEUP), University of Porto, Porto, Portugal

L. F. M. da Silva et al. (eds.), *2nd International Conference on Advanced Joining Processes (AJP 2021)*, Proceedings in Engineering Mechanics,
https://doi.org/10.1007/978-3-030-95463-5_12

differences in behaviour were identified, the proposed bonded joint was found to be suitable for practical implementation.

Keywords Adhesive bonding · Milling tools · Induction cure · Woodworking tools

1 Introduction

Even though a limited number of studies have been carried out since the 1980's on the use of adhesive bonding as a tool joining process, these have been mainly devoted to the assessment of its effects on tool wear and surface quality of metalworking lathe tools, (Darwish 2000a, b; Jachak et al. 2021), or the tool's thermal behaviour during work (Darwish and Davies 1989a; Kilik and Davies 1989; Davies and Darwish 1991; Darwish et al. 1992; Al-Samhan 2012). In the milling tool sector, especially for wood applications, the use of adhesive bonding is still relatively unexplored. Nevertheless, there is a growing interest in the use adhesive bonding as a substitute for the current braze-joined and mechanically fastened processes. With the market's preference for lightweight tools, which require less power to operate and create less stresses in the milling machine, the design of bonded milling tools with light body materials such as aluminium is now a key research subject.

Since induction heating is currently used in the manufacture of brazed tools, one of the main premises of this work was to ensure that this source of heat could also be adopted as the adhesive curing methodology, avoiding the need for costly equipment changes. When used in tungsten carbide/polycrystalline diamond (WC/PCD) inserts—commonly called solely PCD—or steel bodies, this process presents a very high efficiency, reaching the desired temperatures in just a few seconds. Therefore, steel tool prototypes should be able to cure quickly and uniformly under induction heating, since both adherends provide proper heat transfer conditions, as depicted by Fig. 1a.

On the other hand, aluminium—an innovative body material for this type of tools—cannot be inductively heated as efficiently as steel since it does not possess ferromagnetic properties. Consequently, it solely undergoes induction heating via Eddy currents, making the process less effective as a result of the decreased heating

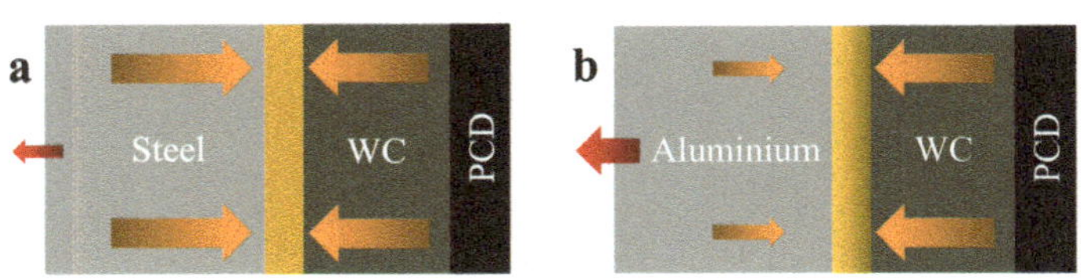

Fig. 1 Schematic representation of the induction cure process in terms of heat fluxes supplied by the coil (*orange arrow*) and diffused by the body (*red arrow*). **a** Steel/PCD uniform cure, **b** Aluminium/PCD nonuniform cure

by Joule effect as well as a quicker heat dissipation due to its higher thermal conductivity. Such conditions create major challenges in the aluminium tool's manufacturing process, leading to non-uniform curing, as it can be seen in Fig. 1b.

Another common concern is the long-term reliability of the bonded tools, partially related to the strong thermal insulation created by the adhesive layer. This was described as a liability for both the adhesive and the insert (Darwish and Davies 1989a, b), due to thermal degradation concerns. With lower heat dissipation through the tool's body, the insert temperature increases when compared with braze-joined or mechanically clamped systems. As a solution, the introduction of conductive elements to improve the thermal conductivity of the adhesive layer was reported by several authors (Kilik and Davies 1989; Davies and Darwish 1991; Al-Samhan 2012).

Since nowadays, most of the adopted insert elements are exceptionally hard, in order to process and sharpen these materials nonabrasive processes such as electric discharge machining (EDM) or laser-based processes are widely used. König said that many different classes of materials such as hardened steels, carbides and ceramics could be processed by EDM if they offered 0.01 S cm^{-1} of electrical conductivity (σ) (König 1991), as shown in Fig. 2.

Therefore, not all materials are sufficiently electrically conductive, especially adhesives, so certain methods are used to overcome this issue, such as adhesive doping. In which by introducing conductive elements as a filler—i.e. aluminium, iron, copper, and graphite powders (Darwish et al. 1991), or conductive spheres (Correia et al. 2021)—proved effective in improving the electrical conductivity of epoxy resins.

Therefore, following the procedure proposed by Correia, (Correia et al. 2021), where an already doped epoxy resin was mixed with conductive spheres, it is possible

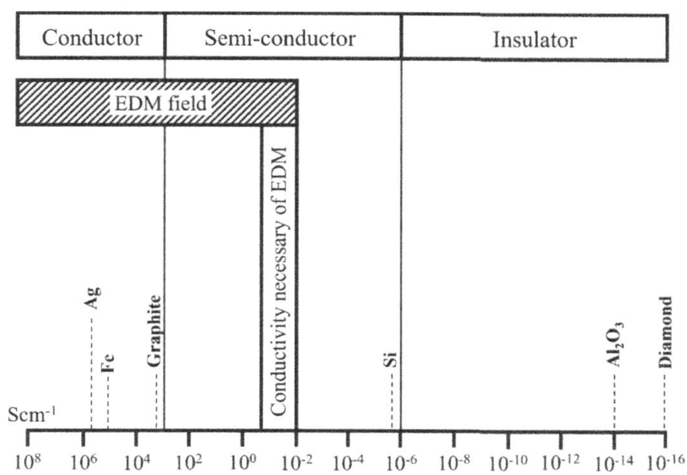

Fig. 2 Electrical conductivity necessary for EDM. (Adapted from König (1991))

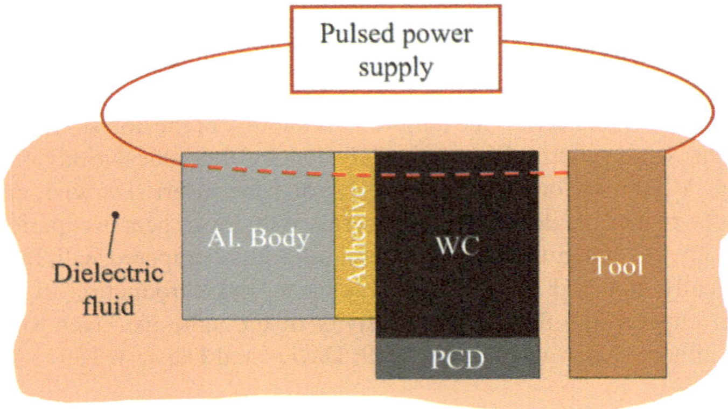

Fig. 3 Schematic representation of EDM sharpening process

to achieve similar layer electrical properties to that of a brazed joint. This promotes the existence of a stable closed loop—Fig. 3—between the body and the tool-electrode, improving the quality of the sharpening process.

The present work consists of a feasibility study on a novel milling tool concept, constructed using PCD inserts bonded to different body materials. This type of bonded construction also has the potential to be used in other applications such as tools with large inserts or geometries highly susceptible to thermal buckling, in which the current processes are difficult to implement or simply not reliable. Aluminium and steel body tools were both manufactured, representing different construction methodologies. Initially, the induction cure cycle was defined and studied using differential scanning calorimetry (DSC) tests. Prototypes were then manufactured for both body material, using an induction curing setup. Finally, the prototypes were properly tested for EDM sharpening and to ensure work safety.

2 Experimental Methodology

The body and insert materials used in this work are part of edge milling tool models already commercially available and were supplied by FREZITE® Woodworking Tools, SA. The adherends treatments, adhesive used, and conductive filler percentages followed the work presented by Correia et al. (2021). The manufacturing and validation procedures were also performed at the partner company.

2.1 Induction Cure Cycle

In order to maintain production rates compatible with an industrial environment, a maximum curing time threshold was set to under a minute. Then two different approaches were established, one seeking to obtain a full cure under the time limit proposed by the company while the other only seeks to obtain a pre-cured state to be finalised in a posterior manufacturing step.

Considering the curing data supplied by the adhesive's manufacturer, the use of a full-cure process is only possible at a temperature—$T_{full\ c}$—higher than 180 °C for the proposed time frame—$t_{full\ c}$. A schematic representation of this cure cycle is presented in Fig. 4.

The pre-cure method can be implemented at lower temperatures—$T_{pre\ c}$—under a similar curing time—$t_{pre\ c}$—since in this process the adhesive is only partially cured to establish enough cohesive strength to maintain the inserts in place in the next cure stage. Afterwards to finalise the cure, a longer time step —t_c—can be established in a convection oven at a lower temperature—T_c. A schematic representation of this cure cycle is presented in Fig. 5.

2.1.1 DSC Tests

Differential scanning calorimetry tests were performed using the adhesive chosen for this study to better understand its polymerisation process at different temperatures and assess the possibility of applying either a quick full-cure or a pre-cure followed by a longer cure stage at lower temperature. All tests were performed at a heating rate of 20 °C min^{-1}.

Two types of tests were performed, a continuous ramp and several isothermal cure stages to determine the curing times at different temperatures. The isothermal

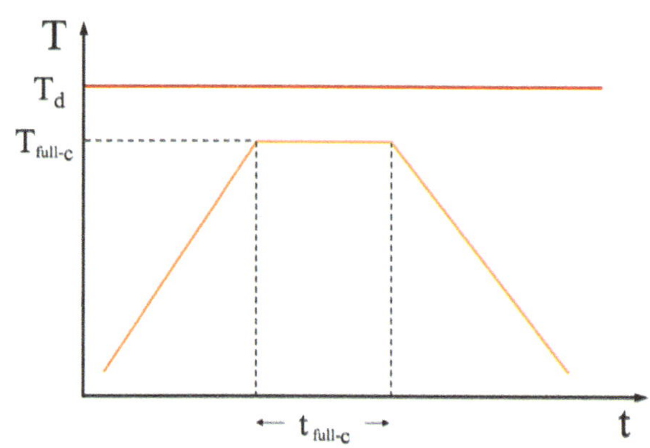

Fig. 4 Full-cure cycle scheme

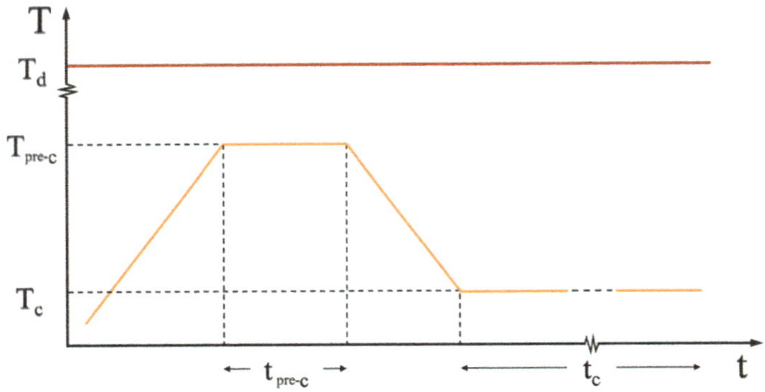

Fig. 5 Pre-cure plus oven cure cycle scheme

cure cycles were performed within the temperatures at which the adhesive reticulation begins and finishes, obtained by analysing the continuous ramp cycle. After reaching the isothermal stage, the cure time was analysed up until the DSC curve stabilises, finalising the adhesive polymerisation process. In the case of the continuous ramp, the heat flow decreases once again until the curves become unstable reaching the degradation temperature—T_d—with the adhesive becoming damaged after this point is reached.

The cure times examined during this study only contemplate the point at which the isothermal stage is reached, ignoring the cycle's heat ramp.

As a result, the adhesive's degradation temperature was accurately determined, safeguarding the use of a damaging cure temperature, and its cure cycles were defined for subsequent practical testing. All sudden irregular curve oscillations should be discarded since they were caused by current peaks.

2.2 *Manufacturing Procedures*

The manufacturing methodology was based on the present induction heating method, repurposing the temperature control unit currently used to control the adhesive's curing cycle instead of the conventional brazing process temperature. The setup used is comprised of a high frequency generator, with a maximum power output of 5.6 kW; a heating head with a water cooled inductor; a foot start/stop switch; a pyrometer with laser assisted positioning and a temperature range of 80–700 °C; and a master control unit, capable of adjusting the isothermal cure temperature—T_c—and settling time—t_c—, the emissivity of the pyrometer, as well as defining the percentage of the maximum power output supplied by the generator, controlling its values based on the temperature readings to maintain the constant temperature stage.

The manufacturing procedure started with surface treatments, where the PCD inserts were degreased in an ultrasound solvent bath, followed by two different stages

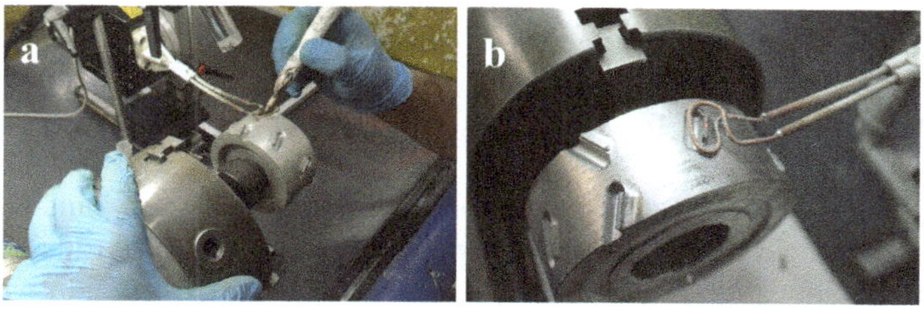

Fig. 6 Induction cured prototype manufacturing. **a** Aluminium body, **b** Steel body

of grit blasting. After this the substrates were cleaned again in an ultrasound bath. All steel bodies were solely grit blasted (St–Gr). As for the aluminium adherends, different paths were tested, following three types of surface treatment configurations. Some of the substrates were only anodised (Al–An), as proposed by Correia et al. (2021), although in this case a sulphuric acid anodisation was selected in order to match current processes used in this industrial environment. However, two other methods were tested during the development of this solution, a grit blasted plus anodised prototype (Al–Gr + An) and a solely grit blasted prototype (Al–Gr).

The positioning of the tool's body was done with a rotational clamping support, to move the insert's groove into position freely. After applying the adhesive, the PCD insert was positioned in place, and cured under appropriate temperature and pressure condition. The temperature induced by the coil was governed by the power supply controller, while pressure was manually applied with a ceramic pointer, both of which can be seen in Fig. 6a and b.

Manufacturing started with a smaller diameter of 100 mm, to assess in a smaller scale the different cure cycle possibilities and to fine tune the induction-based manufacturing procedure. Tools with 180 mm of diameter were later produced, representing the final prototypes to which all the validation methods would be applied.

All prototypes were addressed in the next steps by a code comprised of the prototype's diameter (100 or 180 mm), body material (steel or aluminium) and surface treatments (grit blast or anodization) in the described order, e.g. 180AlGr + An.

2.3 *Work Temperature Influence*

Using a Fluke Ti25 infrared thermal camera, typical work temperature values were determined using a brazed tool while cutting common wood raw materials, such as melamine faced chipboard (MFC) and medium density fibreboard (MDF). Although this test determines only the temperature around the cutting area in a rotating tool, which does not represent the real temperature at the tool cutting edges, it serves as a basis to assess the possibility of temperature degradation issues related to long work stages at severe enough temperatures to damage the properties both the insert and adhesive.

2.4 *Validation Procedures for Woodworking Tools*

2.4.1 EDM Sharpening Tests

Real world sharpening tests were used to evaluate if the bonded joint was able to establish a closed circuit between the workpiece and the tool-electrode performing the EDM sharpening process. This process was implemented by the industrial partner in a commercial EDM machine.

Being performed in a liquid work environment, the possibility of chemical degradation could also be of relevance. However, since the fluid used was a hydrocarbon based dielectric fluid and looking at the chemical medium influence in shear strength under compression detailed in the adhesive's datasheet, chemical degradation should not be a concern since it is unaffected by oils and petroleum-based work mediums for the expected exposure time frame.

2.4.2 Standardised Safety Tests: EN847-1

The safety requirements for wood milling tools are set by the EN 847 standard, Part 1 (2019), where in the specific case of adhesive bonded milling tools, the safety validation methods include a centrifugal test until failure, since in the case of woodworking milling tools, the centrifugal stresses are the most severe ones.

The test specimen should have the largest cutting diameter and width available in catalogue. During the overspeed test, at a rotational test speed of $\eta_p = 2 \cdot \eta$, if no failure is verified in the bonded areas the specimen is considered to have been successfully validated (EN847-1:2017 2019).

The edge milling tool models considered in this work have standard diameters of 100 and 180 mm, with fixed width for each diameter. In terms of rotational velocity, calculations were made considering typical and maximum peripheral cutting velocity's—v_c—common for these tool models, using Eq. 1:

$$\eta = \frac{v_c \cdot 60 \cdot 1000}{\pi \cdot d}, \; [\text{rpm}] \tag{1}$$

whose values are represented in Table 1.

The prototype's overspeed rotational velocity was firstly set to the typical service condition, validating its use, and if no problems arise by that stage, it advances to the next speed level, validating the maximum cutting velocity condition. Since the overspeed testing machine used has a maximum rotational velocity of 21,500 rpm, only the prototypes manufactured with 180 mm can be fully validated at the established rotational velocities.

Table 1 Overspeed values considering two peripheral cutting speed states, for the two standardised diameters used in prototype manufacturing

Speed state	v_c/ms^{-1}	d/mm	η/rpm	η_p/rpm
Typical	70	100	13,000	26,000
		180	7,400	14,800
Maximum	90	100	17,000	34,000
		180	9,500	19,000

3 Experimental Results and Discussion

3.1 Induction Cure Cycle: DSC Test Results

Applying a continuous heating ramp up until 400 °C the DSC plot shown in Fig. 7 was obtained, three main results can be extracted. First, the temperature at which polymerisation begins is slightly lower than 150 °C, the cure process under continuous heating ends at 230 °C, and the degradation temperature is located at 290 °C—as stated in the adhesive's datasheet.

Based on the information obtained from the previous curve, three isothermal tests were carried out at 220, 150 and 180 °C.

Starting with the study of the full-cure cycle, the 220 °C isothermal stage was analysed, its results are presented in Fig. 8. The main conclusion taken from this test was that for this temperature a full cure would not be possible in under a minute. Even less when considering differences in thermal conductivity associated to the aluminium prototypes and possible thermal inertia effects. Though higher cure temperatures could be used, for safety reasons the cure temperature cannot be higher than 230 °C. Furthermore, to achieve these temperatures within the desired time frame, very high heat rates would be necessary, demanding very high-powered equipment and with high likelihood of damaging the adhesive layer and other components of the tool.

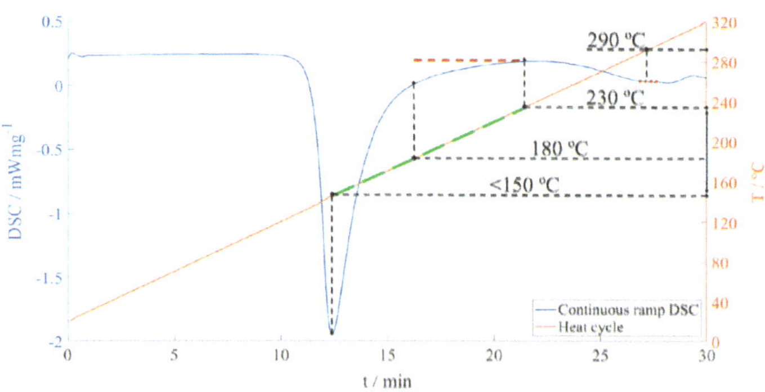

Fig. 7 DSC results for a continuous ramp heating cycle

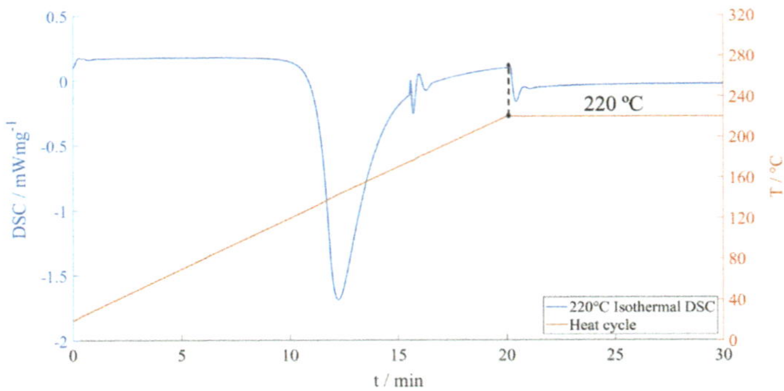

Fig. 8 DSC results for an isothermal heating cycle at 220 °C

Thus, the pursue of a full-cure cycle was abandoned going forward. Nevertheless, this analysis determined that a quick pre-cure cycle at this temperature could be successfully established.

For the pre-cure cycle study the two other isothermal tests were analysed, at 150 and 180 °C, whose results are presented in Figs. 9 and 10, respectively.

Considering the 150 °C stage, it is possible to extrapolate from the DSC curve that more than three minutes are necessary to obtain a suitable pre-cured state, which is not compatible with the manufacturing time frame previously established. Consequentially, this temperature was removed from consideration.

As for the cure stage at 180 °C, contrarily to the previous test, when the isothermal temperature is reached the curve is almost stabilized, which make this cure cycle suitable for pre-curing.

All results considered, the full cure cycle was discarded, and a few pre-cure cycles were selected for testing in the manufacturing stage, namely the 180 and 220 °C pre-cure cycles.

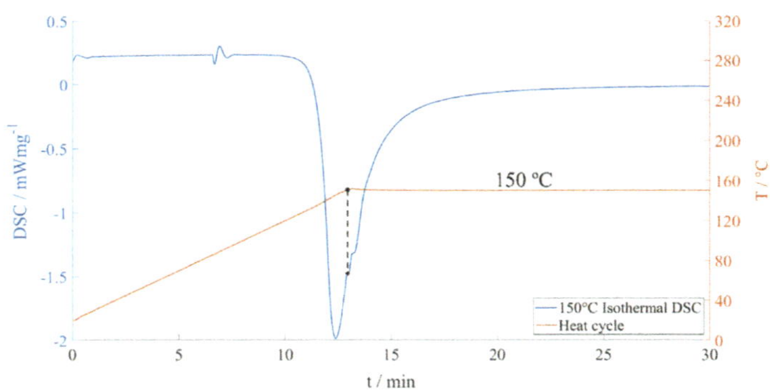

Fig. 9 DSC results for an isothermal heating cycle at 150 °C

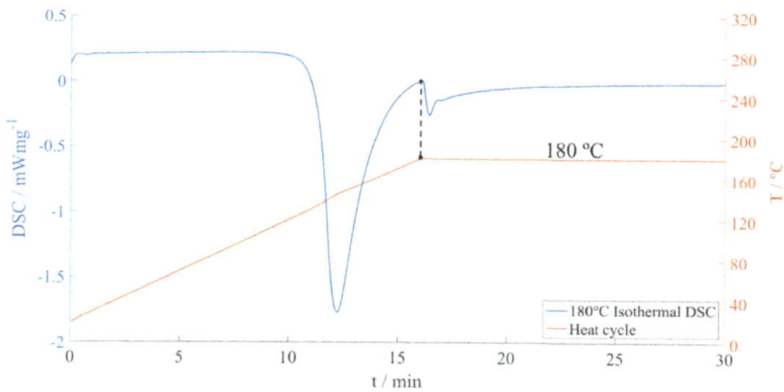

Fig. 10 DSC results for an isothermal heating cycle at 180 °C

3.2 Prototype Manufacturing

For the 100St–Gr prototypes, after testing all pre-cure cycles, it was possible to access the need for a slow heat ramp, considering the lowest power output possible, to prevent a severe expansion effect. This effect was found for the 220 °C cure temperature, making so that the only cure cycle possible for steel tools was the gentler isothermal stage at 180 °C, with low power output. Now, not showing any manufacturing concerns for the smaller diameter a 180St–Gr prototype was manufactured with the same cure cycle for the final validation tests.

When performing the same tests for the 100Al–An, 100Al–Gr + An and 100Al–Gr prototype configurations the same did not occur, mainly due to the low induction heating efficiency and high thermal diffusivity of the aluminium body. In this case, for all cycles a higher power output was necessary to improve the pre-cure performance, however, the previously detected expansion effect did not appear in any of the cure stages. This said, the selected cure cycle would simply translate into a specific curing time. Because productivity is one of the main requirements of this study, to the fastest cycle—220 °C isothermal cure stage—a few extra steps were taken to improve the manufacturing time of these prototypes. Since the 100 mm prototypes showed curing issues for the current adopted manufacturing process, no 180 mm diameter aluminium tool configurations were manufactured, as the steps taken to improve the curing process on the aluminium adherend did not work either properly nor efficiently for the highest diameter.

The final prototype configurations finished their curing process through a longer cure stage at lower temperatures, achieving the state—Fig. 11—where all bonded joints are fully cured. Finally, they proceeded to the validation stage where the following configurations were tested: 100Al–An, 100Al–Gr + An, 100Al–Gr and 180St–Gr.

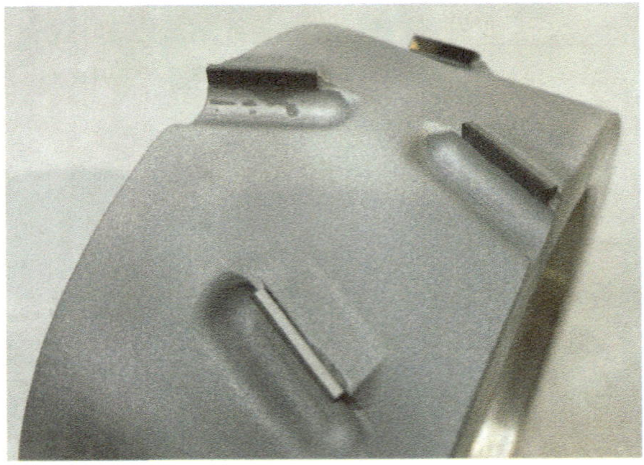

Fig. 11 Bonded joints of prototype 100Al–Gr

3.3 *Work Temperature Measurements*

The work temperature values measured using an infrared thermal camera showed low temperature readings, presented in Fig. 12.

Taking into consideration that in this application the use of hard materials with very low friction coefficients—such as PCD inserts—to decrease cutting edge temperatures, coupled with aluminium bodies with much higher thermal conductivity. But also having a rotational velocity that promotes heat dissipation by means of convection, and softer workpiece materials that produce less heat during chip creation, it is possible to safely consider that thermal degradation of both the insert and adhesive will not be a relevant factor in woodworking adhesively bonded milling tools.

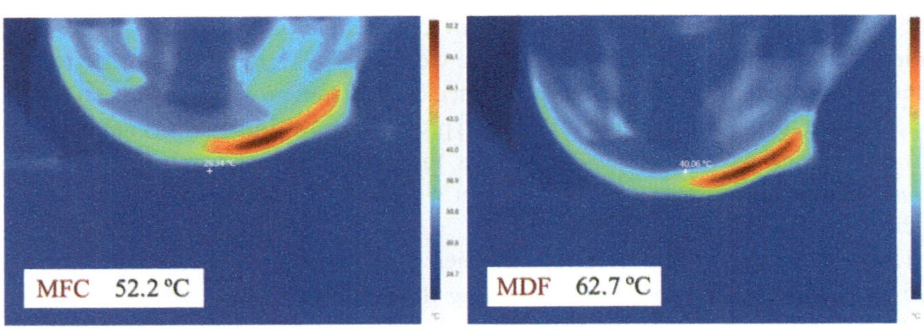

Fig. 12 Thermal imagery of two woodworking scenarios, MFC to the left and MDF to the right

Table 2 Summary of the validation test results for each prototype

Prototype	EDM test	Overspeed test		Joint failure
	Y/N	Typ./η_p %	Max./η_p %	Y/N
100Al–An	N	82	63	N
100Al–Gr + An	N	82	63	N
100Al–Gr	Y	82	63	N
180Al–An	Y	100	100	N

3.4 Woodworking Tool Validation Tests

A summary of all the tests performed, both for EDM and work safety, is presented in Table 2. The sharpening tests are evaluated on a Yes/No basis, and the safety requirements on the test velocity achieved in relation to each overspeed condition, as a percentage. All this to evaluate the results taking into consideration the test machine limitations, as well as the occurrence of joint failure evaluated with a binary Yes/No criterion.

Both aluminium anodised prototypes—100Al–An and 100Al–Gr + An—failed in the sharpening tests due to the insulation effect of the present anodisation layer, which for the previous bonding study (Correia et al. 2021), did not present a problem when measuring the adhesive layer's resistance. This might have occurred since different anodisation processes were used in each study, possibly changing the amount of influence that the anodisation layer had on the electrical properties. Nevertheless, both the 100Al–Gr and 180St–Gr prototypes were successfully sharpened via EDM, validating the use of this adhesive configuration in terms of electrical conductivity in industrial conditions.

From all the aluminium prototypes manufactured, with 100 mm in diameter, none were tested for the predicted overspeed conditions due to the machine's speed limitation. Nevertheless, all achieved the maximum speed supported by the machine without failure, corresponding to a rotational speed safety factor of 1.65 and 1.26 for the typical (26,000 rpm) and maximum (34,000 rpm) states, respectively. Values slightly smaller than the coefficient of safety of 2.00 required by the standard. As for the 180St–Gr prototype all overspeed states were achieved without any sign of fracture, both in typical (14,800 rpm) and maximum (19,000 rpm) conditions, fully validating this configuration for the safety requirements set by the EN 847 standard.

4 Conclusions

The main purpose of this work was to validate the application of adhesive bonding in the production of milling tools. A suitable induction cure cycle was studied, first in a laboratorial setting with DSC analysis and then practically during manufacturing of the prototype configurations. Both steel and aluminium solutions were tested having different surface treatments. EDM sharpening tests were performed as well as overspeed tests in accordance with the EN 847 standard. This led to the following findings:

- Unlike the brazing process, bonded tools require a two-step cure cycle starting with a quick pre-cured followed by a longer oven cure;
- Neither chemical nor thermal degradation seem problematic for both the production and work stages of wood milling tools;
- Aluminium tool bodies, contrarily to steel, posed a few challenges when curing the adhesive under an induction based process;
- The doped adhesive performed as expected, increasing the electrical conductivity of the adhesive layer, allowing EDM sharpening of the grit blasted prototypes;
- The anodisation of the aluminium precludes EDM sharpening and thus only laser based sharpening solutions can be implemented with this surface treatment;
- The only prototype that was fully validated was the steel configuration, showing that a few improvements in the manufacturing process are still needed to bond aluminium tools;
- Other adhesive fillers or pre-curing solutions might improve the productivity of this manufacturing process.

Although important differences in behaviour between brazed, steel bonded and aluminium bonded joints were identified, the proposed bonded tools were found to be suitable for practical implementation, even though in the case of aluminium a few challenges must still be overcome. Nevertheless, this is a relatively incipient process, with plenty of room for achieving greater efficiency, performance, and productivity in an industrial environment.

References

Al-Samhan, A.M.: Thermal-stresses in carbide-tip bonded face milling cutters. J. King Saud Univ. Eng. Sci. **24**(2), 85–94 (2012)

Correia, D.S., Carbas, R.J.C., Marques, E.A.S., das Neves, P.J.C., da Silva, L.F.M.: Experimental study on aluminium to WC/PCD adhesive bonding for milling tools. Under Revision on Int. J. Adhes. Adhes. (2021)

Darwish, S.: Effect of tool bit insert-holder assembly on the quality of machined workpieces. J. Mater. Process. Technol. **105**(3), 230–236 (2000a)

Darwish, S.: Machining of difficult-to-cut materials with bonded tools. Int. J. Adhes. Adhes. **20**(4), 279–289 (2000b)

Darwish, S., Davies, R.: Investigation of the heat flow through bonded and brazed metal cutting tools. Int. J. Mach. Tools Manuf. **29**(2), 229–237 (1989a)

Darwish, S., Davies, R.: Adhesive bonding of metal cutting tools. Int. J. Mach. Tools Manuf. **29**(1), 141–152 (1989b)

Darwish, S., Niazi, A., Ghania, A., Kassem, M.E.: Improving the electrical properties of structural epoxy resin adhesives. Int. J. Adhes. Adhes. **11**(1), 37–42 (1991)

Darwish, S., Niazi, A., Ghaneya, A.: Phase stability of duralumin machined with bonded and brazed carbide tools. Int. J. Mach. Tools Manuf. **32**(4), 593–600 (1992)

Davies, R., Darwish, S.: Temperature distribution of bonded tools having atomized copper powder mixed with adhesive. Int. J. Mach. Tools Manuf. **31**(4), 617–623 (1991)

EN847-1:2017: Tools for woodworking—safety requirements—Part 1: milling tools, circular saw blades. European Committee for Standardization (2019)

Jachak, S., Giri, J., Awari, G.K., Bonde, A.S.: Surface finish generated in turning of medium carbon steel parts using conventional and adhesive bonded tools. Mater. Today Proc. **43**, 2882–2887 (2021)

Kilik, R., Davies, R.: Mechanical properties of adhesive filled with metal powders. Int. J. Adhes. Adhes. **9**(4), 224–228 (1989)

König, W.: Advanced ceramics: sparks machine ceramics. Powder Metal. Int. **23**(2), 96–100 (1991)

Correction to: Practical Implementation and Validation of a Novel Process for Manufacturing Milling Tools Using Adhesive Bonding

D. S. Correia, E. A. S. Marques, R. J. C. Carbas, P. J. C. das Neves, and L. F. M. da Silva

Correction to:
Chapter "Practical Implementation and Validation of a Novel Process for Manufacturing Milling Tools Using Adhesive Bonding" in: L. F. M. da Silva et al. (eds.), *2nd International Conference on Advanced Joining Processes (AJP 2021)*, Proceedings in Engineering Mechanics https://doi.org/10.1007/978-3-030-95463-5_12

The original version of this chapter was published with incorrect author names in XML files and hence they were incorrectly cited online. This has now been rectified and the author names have been updated in XML.

The chapter and book have been updated with the changes.

The updated version of this chapter can be found at
https://doi.org/10.1007/978-3-030-95463-5_12

L. F. M. da Silva et al. (eds.), *2nd International Conference on Advanced Joining Processes (AJP 2021)*, Proceedings in Engineering Mechanics,
https://doi.org/10.1007/978-3-030-95463-5_13

The manufacturer's authorised representative in the EU is Springer Nature Customer Service Centre GmbH, Europaplatz 3, 69115 Heidelberg, Germany. If you have any concerns regarding our products, please contact ProductSafety@springernature.com

Printed and bound by CPI Group (UK) Ltd, Croydon, CR0 4YY
30/09/2026
02225324-0003